LEAN SIX SIGMA FOR BEGINNERS

Essential Activities to Conduct DMAIC Projects

Gary Jing

QUALITY PRESS

Milwaukee, Wisconsin

American Society for Quality, Quality Press, Milwaukee 53203
© 2025 by Quality Press.
All rights reserved. Published 2025
Printed in the United States of America

29 28 27 26 25 LS 5 4 3 2 1

Publisher's Cataloging-in-Publication data

Names: Jing, Gary G., 1964-, author.
Title: Lean six sigma for beginners: essential activities to conduct
  dmaic projects / by Gary Jing.
Series: ASQ's Pocket Guide Series
Description: Includes bibliographical references. | Milwaukee, WI:
  Quality Press, 2025.
Identifiers: LCCN: 2025930172 | ISBN: 9781636941882 (paperback) |
  9781636941899 (PDF) | 9781636941905 (epub)
Subjects: LCSH Six sigma (Quality control standard) | Project
  management. | Process control. | Inventory control. | Industrial
  efficiency. | Cost effectiveness. | BISAC BUSINESS & ECONOMICS /
  Quality Control | BUSINESS & ECONOMICS / Industries /
  Manufacturing
Classification: LCC HD69.P75 .J56 2025 | DDC 658.4/013—dc23

ASQ advances individual, organizational, and community excellence
worldwide through learning, quality improvement, and knowledge
exchange.

Bookstores, wholesalers, schools, libraries, businesses, and organizations:
Quality Press books are available at quantity discounts for bulk purchases
for business, trade, or educational uses. For more information, please contact
Quality Press at 800-248-1946 or ask@asq.org.

To place orders or browse the selection of all Quality Press titles,
visit our website at: http://www.asq.org/quality-press

Printed in the United States of America.

Quality Press
600 N. Plankinton Ave.
Milwaukee, WI 53203-2914
Email: books@asq.org
QUALITY
PRESS Excellence Through Quality™

Table of Contents

Improve Phase

Control Phase

List of Figures and Tables

Preface

Within this book, you'll find best practices for Lean Six Sigma (LSS). The book is designed to quickly and easily equip young professionals and college students with practical LSS skills for real-world use. It's based on the hands-on, results-driven curriculum I use in the Lean Six Sigma course at the University of St. Thomas. Half of these LSS students are working young professionals; thus, the book is aimed at and structured for the needs and learning styles of young professionals.

You may:

- Scan to distill and learn at the point of need, or
- Leverage free resources in the public domain to support just-in-time learning.

This book:

- Focuses on "soft," or non-statistical LSS tools, with the assumption that young professionals are better prepared to work with data analytics and statistics but are less experienced with handling "soft" activities.

- Follows a typical project roadmap that users can easily follow, step-by-step, to carry out LSS projects.

- Offers some unique and lesser-known practices and perspectives on popular tools, such as root cause analysis (RCA) and failure mode and effects analysis (FMEA).

- Develops next-generation quality professionals to be more job competitive by preparing and assisting in the student-to-professional transition and distilling topics down for easy scanning.

SCOPE OF THE BOOK

This book focuses on Define-Measure-Analyze-Improve-Control (DMAIC) Yellow Belt (YB) content and light LSS Green Belt (GB) content, as shown in Figure 0.1, where hours refer to the training associated with each level of progression when Six Sigma (SS) was originally created. DMAIC and Lean Design for Six Sigma (LDFSS) represent typical LSS frameworks.

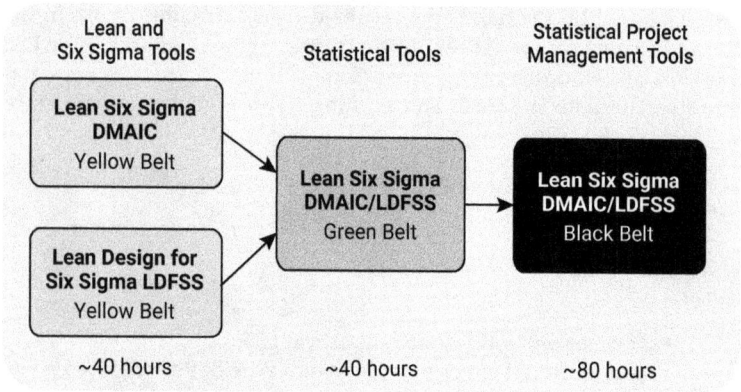

Figure 0.1 LSS tier example.

There are two types of LSS certifications:

1. Exam-focused, which is typically used by institutions (third parties)

2. Project-focused, which is typically used by employers

This book is more practice- or project-focused, although it can be used to prepare for certification exams.

STRUCTURE OF THE BOOK

This book follows a typical LSS project roadmap, addressing both high-level phases and sub-level steps in a progressional way:

- Chapter 1 introduces Six Sigma, lean, and LSS.

- Chapters 2–4 cover the typical activities performed in the Define (D) phase.

- Chapters 5–8 focus on the typical "soft" (non-statistical) activities tackled in the Measure (M) and Analyze (A) phases.

- Chapters 9–15 include signature data analytics used by LSS projects that are typically completed in the Analyze phase.

- Chapters 16–22 explain the popular lean tools used by LSS projects in the Improve (I) phase.

- Chapters 23–24 cover the typical activities completed in the Control (C) phase.

Note that although DMAIC phases are presented sequentially, in practice they tend to be clustered in two main stages: DMA and IC, with many iteration loops within each stage. Many tools and/or applications have a primary phase in which they are used, but they may also be used in other phases throughout the project. In this book, I'll cover the non-statistical analyses in the Measure phase and statistical analyses in the Analyze phase, although they may be used in both phases. For the same reason, I will dedicate the Improve phase to lean tools and applications, where they are more likely used—although some lean activities may also occur in other phases.

For additional resources to accompany this book, contact the author at https://www.linkedin.com/in/ggaryjing/.

Acknowledgments

I'd like to thank Mary McShane-Vaughn for mentoring me during the writing of this book and for giving me extremely valuable feedback and suggestions.

CONTRIBUTORS' ACKNOWLEDGMENTS

The American Society for Quality and Quality Press would like to thank the Quality Press Peer Review Committee for its invaluable volunteer participation and contributions to this work. Without our volunteers' subject matter expertise, time, and passion for creating content, none of our efforts would be possible.

Quality Press Peer Review Committee Members

Scott A. Laman, Chair
Peter Pylipow, Vice Chair
Melvin Alexander
Lance Coleman
Ahmad Elshennawy
Marc Hamilton
Gary Jing
Trevor Jordan
Jane Keathley
Mary McShane-Vaughn
Jayet Moon
G.S. Sureshchandar
Tiea Theurer

Chapter 1
Introduction to Six Sigma

Six Sigma (SS) was developed in the mid-1980s by Motorola and popularized by General Electric (GE). Motorola sought a systematic way to improve quality and performance, which led to SS, a data-driven approach aimed at achieving near-perfect quality by identifying and eliminating defects.

SIX SIGMA OVERVIEW

Six Sigma is a metrics, methodology, and management philosophy and strategy, as described below.

- *Measurement and metric:* A process with ± six standard deviations (σ) from the mean (μ) falling within specifications.

- *Methodology* (to make improvements or solve problems)

  - DMAIC framework: The iconic structure of SS projects stands for Define, Measure, Analyze, Improve, and Control.

  - Focus on variation reduction: SS emphasizes minimizing variation to ensure consistent and predictable outcomes.

  - Data-driven decision-making: Decisions are based on rigorous data analysis, fostering objective and fact-based approaches to problem solving.

  - Customer focus: SS prioritizes meeting customer requirements and enhancing customer satisfaction by delivering high-quality products and services.

- *Management philosophy, business initiative, and strategy*
  - Right projects: Breakthrough improvements linked to business goals, focusing on reducing variability and errors/defects
  - Right people: SS belts trained with SS capability
  - Right results: Sustained financial gains

SIX SIGMA AS A METHODOLOGY

There are two types of SS roadmaps:

1. DMAIC (aka *operational SS*), a disciplined framework used to improve an existing product or process. Although phases and training are sequential, in practice they are more clustered into two main stages—DMA and IC—with many iteration loops within each stage, as shown in Figure 1.1. This book follows the DMAIC phase structure, the original and signatory SS framework.

2. *Design for Six Sigma* (DFSS), the disciplined framework for designing a new product or process. Unlike DMAIC as a commonly adopted structure, DFSS has different structures that are adopted in various situations, such as CDOV/IDOV (Concept/Ideation, Design, Optimization, and Verification) or DMADV (Define, Measure, Analyze, Design, Validate).

Typical SS tools include:

- Qualitative tools, such as brainstorming, process mapping,* root cause analysis (RCA),* failure mode and effects analysis (FMEA),* and control plan.*

- Quantitative tools, such as check sheets, histograms, Pareto analysis, statistical process control (SPC),** measurement system analysis (MSA),** capability analysis,** and design of experiments (DOE).

---

  * Core DMAIC applications.
 ** Signature Six Sigma data analytics (emphasized by SS but not by other approaches).

Define	Measure	Analyze	Improve	Control
• Identify the gap • Assign belt and establish scope and boundary • Determine customer requirements • Assign team • Establish the project charter	• Process maps*/value stream maps (VSM) • Determine key process input variables (KPIVs), key process output variables (KPOVs) • RCA* • FMEA*	• Graphical analyses • Stability (SPC)** • MSA** • Capability analysis** • Hypothesis testing • Correlation and regression	• Process optimization (lean) • DOE • Pilot new process • Confirm results • Update VSM	• Implement error proofing • Implement SPC • Verify long-term capability • Implement control plan* • Update FMEA • Return to process owner

Figure 1.1　A DMAIC roadmap.

SIX SIGMA VS. OTHER APPROACHES (PDCA, 8D, A3)

There are three other approaches that are comparable with SS. They are: PDCA, 8D, and A3. PDCA refers to the plan, do, check, act improvement cycle. 8D refers to the eight-discipline problem-solving approach, as shown in Table 1.1. A3 refers to the practice Toyota developed to document the problem, the analysis, the corrective actions, and the action plan on an A3 (11 x 17-inch) paper. All can do the job, each with its own flavor, although SS has the strongest and deepest supporting infrastructure. It has the most extensive training with progressive tiers of capability designation, thus geared toward more sophisticated, deep-dive types of situations or applications. The levels and associated training when SS was originally created include:

- Yellow Belt (YB): one week (~40 hours) of training
- Green Belt (GB): two weeks (~80 hours) of training
- Black Belt (BB): four weeks (~160 hours) of training
- Master Black Belt (MBB): six weeks (~240 hours) of training

Table 1.1   DMAIC vs. alternative models comparison.

PDCA	DMAIC	8D	A3
Plan	Define	1. Create a team	Clarify the problem
	Measure	2. Describe the problem	Understand the current state
		3. Define containment action	Analyze the root cause
	Analyze	4. Analyze the root cause	Develop a countermeasure
		5. Define a corrective action	Define the target state
Do	Improve	6. Implement the corrective action	Implement the countermeasure
Check	Control	7. Take action to prevent recurrence	Evaluate the result and process
Act		8. Congratulate the team	Standardize

The rest of the approaches have at most a couple of days of training, which is geared toward more straightforward situations or applications. In a way, a lot of the differences are apparent in how the story is structured and presented. While SS is more sophisticated, it can be a double-edged sword, as SS also has an efficiency challenge. People often see it as inefficient, burdensome, or overkill for a given situation.

LEAN OVERVIEW

The term "lean" was coined in the late 1980s and popularized by the book *The Machine That Changed the World* (Womack et al., 2007). Lean is used to highlight the principles and practices of the famous Toyota Production System (TPS), which mastered the trade of improving productivity and efficiency through continuous improvement (kaizen) and waste elimination.

Lean Principles

- Identifying value: Specify what is important to the customer.
- Mapping the value stream: Identify the activities that contribute to customer value.
- Creating flow: Build a process with value-added steps processing through the value stream.
- Establishing a pull system: Ensure your system is driven by customer demand.
- Seeking perfection: Continuously inspect your system for signs of waste and eliminate them.

Value-Added and Value Stream

The *value* is defined from the customer's perspective, representing any activity, resource, or input that directly addresses the needs of the customer. For example, while assembling subcomponents adds value, walking to get subassemblies does not. Correspondingly, there are three types of activities:

1. Value-added (VA)

    - It transforms or shapes raw material or information.
    - It must be done right the first time.
    - The customer values (is willing to pay for) the output of the activity.

2. Non-value-added (NVA): It doesn't meet the definition of VA and is removable.

3. Non-value-added but necessary: It doesn't meet the definition of VA but is required by the business or other regulatory bodies.

A *value stream* is an end-to-end chain of process steps that starts from a customer trigger and proceeds until the customer's need is successfully fulfilled. A *value stream map* (VSM) is a mapping format used in lean to depict a value stream.

Eight Wastes (Muda)

This is a core concept of lean, frequently referred to as DOWNTIME:

- Defects: products or services that do not conform to specifications—any rework or scrap

- Overproduction: producing more, earlier, or faster than what's required by the next process

- Waiting: any idle time or delay waiting for materials, machines, people, service, etc.

- Non-utilized talent (skills): underutilizing peoples' talents, skills, and knowledge

- Transportation: any movement of material that is nonessential

- Inventory: products or services more than the minimum required to complete the job

- Motion: unnecessary movements by people

- Extra/over-processing: doing more than necessary to meet requirements

Eight wastes is also sometimes called TIMWOODS: Transportation, Inventory, Motion, Waiting, Overprocessing, Overproduction, Defects, Skill underutilized.

Lean Strategies

Eliminate waste and create flow across the entire value stream, which includes:

- Driving small lot sizes, mixed production, and pull from the customer

- Arranging plants according to process flow

- Working to prevent errors and eliminate waste

- Improving delivery performance by having short cycle times with minimal work-in-process and finished goods inventory

Popular Lean Tools

Lean tools or techniques are most used in the Improve phase. Determining which tools to use is situational, depending on the need.

- Fundamental: Voice of the customer (VOC), diagnostic process mapping (DPM)/VSM, kaizen, 5S + visual workplace, standard work (SW), or mistake proofing (MP)

- Advanced: Continuous flow (CF), setup reduction (SR), cell design (CD), or production preparation process (3P)

LEAN AND SIX SIGMA COMPARISON

Lean and SS complement each other, yet each has its own strengths and limitations. Lean and Six Sigma look at the same things from two different angles, intertwining with each other.

- Lean focuses on cycle time and throughput via eliminating waste; it is biased toward action and is used more for efficiency-related problems.

- Six Sigma aims to achieve consistent output by identifying and reducing variation and defects; it is biased toward analysis and is used more for conformance-related problems.

A problem usually can be addressed from either or both sides and, rarely, neither. Changes or solutions induced from one side will likely affect the other side; thus, there could also be friction or conflict on the ground. Some lean practitioners don't see the need for SS, yet SS practitioners usually see the need for lean.

LEAN SIX SIGMA (LSS) INTEGRATION

Although friction or conflicts do exist, they are overwhelmingly outweighed by synergies; thus, nowadays lean and SS are more frequently seen as LSS integrated, either in progressional tiers or following SS frameworks such as DMAIC. A progressional-tier example would be: 5S → Kaizen → Lean → Sigma → DFSS. Lean activities can follow the DMAIC framework if you choose.

Key Takeaways

- Six Sigma is a measurement and metric, a methodology to make improvements or solve problems, and a management philosophy, business initiative, and strategy.

- Lean originated from TPS and focuses on maximizing value by eliminating waste and improving efficiency.

- Lean principles include identifying value, mapping the value stream, creating flow, establishing a pull system, and seeking perfection.

- Three things are required for an activity to be value-added: a transformation of the material or information, doing it right the first time, and having customers who are willing to pay.

- The eight wastes (Muda) help identify areas of opportunities— TIMWOODS: Transportation, Inventory, Motion, Waiting, Overprocessing, Overproduction, Defects, Skill underutilized.

- Popular lean tools include VSM, kaizen, 5S + visual workplace, standard work, mistake proofing, continuous flow, setup reduction (SR), and cell design.

- LSS combines the strengths of lean (efficiency) and Six Sigma (quality) into a comprehensive improvement methodology.

- Friction or conflict do exist but are outweighed by synergy, which provides the holistic approach of LSS to problem solving, combining complementary tools and techniques.

- The LSS methodology fosters a culture of continuous improvement supported by data-driven decision-making.

Define Phase

The focus of the Define phase is to identify the most rewarding opportunity in need of attention. In this phase, you will define the opportunity—as concretely as possible—to lay a solid foundation for the project to succeed.

There was no Define phase when SS was originally developed. Practitioners learned the hard way that projects stumble a lot without sound investment in project definition; thus, the Define phase was added.

Chapter 2
Project Mining and Selection

Having established a foundational understanding of LSS, we now move to the first practical step in applying the methodology: *project mining and selection*. This chapter is essential at the beginning of LSS activities because to effectively apply LSS, you must identify and prioritize the right high-impact opportunities to ensure your efforts are aligned with strategic goals and yield significant improvements. Project selection typically involves three steps: mining, scoping, and prioritization.

PROJECT MINING

Identify your basic Six Sigma project.

What Constitutes a Sigma "Project?"

- A gap exists between current and required performance.
- The causes of the problem are not clearly understood.
- The solution is not predetermined nor is the optimal solution apparent.

People sometimes argue whether an improvement is a kaizen or Six Sigma project. Here are some guidelines:

- If it has the signature Six Sigma activities/applications, it will constitute a Six Sigma project; for example, it may have a DMAIC structure with five core applications: process mapping, RCA, FMEA, data analytics, and control plan.

- An improvement can be both a kaizen (small, incremental change) and a Six Sigma project. It's less about the type of improvement and more about the applications/activities (methodology), which may just be personal preference.

- A good practice is to integrate. Add SS applications to kaizen, or expand some kaizen into Six Sigma projects, and you've got a one-two punch.

Project Ideas Generation

This typically involves three steps:

1. Brainstorm daily challenges. Generate a list of currently important issues/problems. VSM or process maps are frequently used to identify gaps between the actual and the intended results.

2. Identify the scopes and venues.

3. Prioritize opportunities via voting or scoring.

PROJECT SCOPING

The following techniques can help scope the project:

- Be *SMART*: Specific, Measurable, Attainable, Realistic, and Time-bound.

- Understand the span-of-control/sphere-of-influence (see Figure 2.1):

   - Span of control *(ideal scope)* = the innermost layer of the three-layer circles, where one has full authority and there is high impact and low cost

   - Sphere of influence *(acceptable scope)* = the encompassing middle layer where one has influence and there is medium impact and medium cost

   - Out of reach *(unacceptable scope)* = the outer circle where one has no influence or control and there is low impact and high cost

- Use a *root-cause tree*. Drill down to a level that:

  - Is manageable and measurable

  - Has clear process owners and access to resources

  - Has no significant or obvious interactions or interdependencies at this level

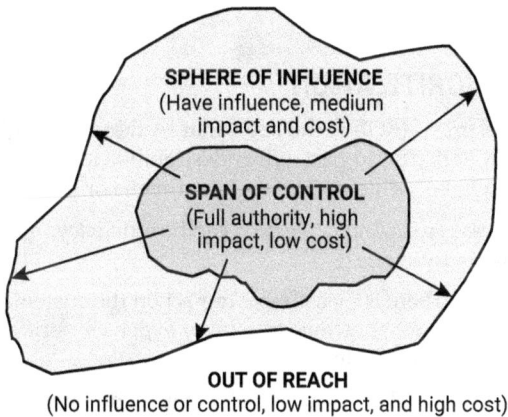

Figure 2.1 Span of control/sphere of influence.

An LSS project should be SMART, preferably within the span of control or at least within the sphere of influence, and have no significant or obvious interactions or interdependencies with other projects.

Below are some notes on the various types of projects:

- Kaizen events are small and tactical (with a narrow scope) projects with a short project cycle (less than one week, with a team that is 100 percent dedicated). Kaizen events can be part of larger BB/GB projects, which may use kaizen as a vehicle for Improve phase activities.

- Yellow Belt/Green Belt projects are small or tactical projects with a medium project cycle (one to three months, led by YB/GB, with a team that is not 100 percent dedicated. Events are part of the daily tasks instead).

- Black Belt/Master Black Belt projects are large or strategic projects (three to six months, led by a BB or MBB, with a team that is not 100 percent dedicated. Events are part of daily tasks).

PROJECT PRIORITIZATION

Return on investment (ROI) is the key—that is, there is a high priority placed on high ROI. The impact–effort desirability matrix (see Figure 2.2) is a good ROI tool, which evaluates two dimensions:

- Business return/impact, which can be efficiency, productivity, etc.

  – High = There is measurable impact on the customer, business or net income (capital or expense), verifiable safety improvements, etc.

  – Medium = There is measurable impact on internal metrics or cost avoidance.

  – Low = Impact is not measurable but moves in a direction of goodness.

- Difficulty/effort/cost, which can be measured by complexity, dollars invested, resources required, etc.

  – High = requiring significant time, capital, expense, or outside resources

  – Medium = requiring moderate time and/or internal resources outside of immediate function

  – Low = directly implementable by the project owner

The desirability of a project increases as you move from the lower right to the upper left, and as the circle gets larger

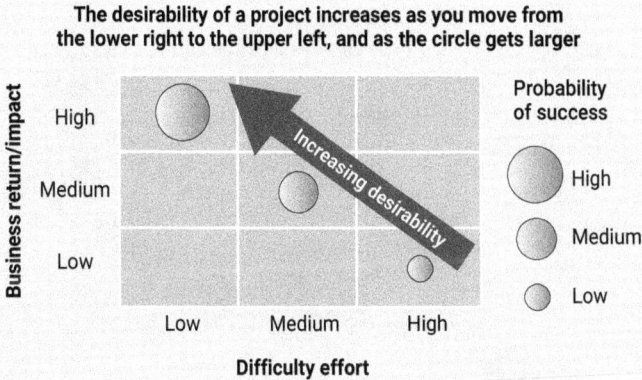

Figure 2.2 Desirability matrix (example).

Notice the ROI calculation here doesn't have to be the one used for finance, which is the net return ($)/cost ($). For prioritization purposes, it can be any form of assessed return/assessed cost. Another good ROI tool is the cause-effect matrix, where projects are listed on the *cause* side and evaluation criteria are listed on the *effect* side. Refer to Chapter 7 for details.

Key Takeaways

- Project mining and selection are essential to ensure LSS efforts are focused on high-impact opportunities.

- Project selection typically involves three steps: mining, scoping, and prioritization.

- Identifying opportunities involves analyzing data, engaging stakeholders, and aligning with strategic goals to identify gaps.

- An LSS project should be SMART, preferably within your span of control or at least within your sphere of influence, and have no significant or obvious interactions or interdependencies with other projects.

- Projects should be prioritized based on ROI and/or the impact-effort desirability matrix.

Chapter 3
Project Definition

Having identified and selected high-impact projects, we now turn to defining the project with clarity and precision. This is a critical step to ensure project success. This chapter discusses the key elements of the project charter to prepare a sound footing for project success.

PROBLEM STATEMENT

A good problem statement contains:

- WHAT is wrong (the *defect*)
- HOW it is wrong (to *quantify*)
- WHERE and WHEN does it happen
- WHY it is a problem (business impact)

It can mention what's critical to quality, delivery, or something (CTQ, CTD, CTX). It should *not* mention any causes or solutions. A defect is a shortcoming, imperfection, or nonconformance against a target or specifications.

Problem statement examples are listed below.

- A **poor** example: *Product returns are too high due to product A and will be reduced by analyzing the first- and second-level Pareto charts.*
- A **good** example: *Product returns are 5 percent of sales, resulting in a profit impact of $5M and customer dissatisfaction.*

OBJECTIVE STATEMENT

Follow the SMART goal: be Specific, Measurable, Attainable, Relevant, and Time-bound.

- Template: To increase/decrease the {metric or KPOV (key process output variable) } from {baseline level} to {goal level} by {time frame} as measured by {measurement system}

- Example: *To reduce the actual cycle time by 5.93 days overall (approximately 14 percent of current) within two fiscal quarters.*

Scope and Stakeholder

- Define the boundaries of the project: what's included and excluded. A clear scope definition helps prevent scope creep.

- Identify all stakeholders involved in or affected by the project. Understanding their needs and expectations is crucial for project success.

Primary and Secondary Metrics, Business Benefit

Define primary and secondary metrics:

- The primary metric is the yardstick to measure success (for example, cycle time, defect rate, etc.). It must be connected to the problem statement and objective directly.

- The secondary (consequential) metric is the conscience to "keep you honest." It tracks potentially negative consequences of achieving the primary metric—for example, headcount, customer satisfaction, etc.

Business benefits must identify at least one of the CTX and link to the business goals. Estimate hard and soft savings:

- *Hard savings* are directly recognizable by accounting and/ or reflected in the budget/P&L; for example, this could be reduced scrap, rework, workforce, overtime, or operating expenses (supplies, utility usage, equipment depreciation, lease or rent, etc.). Cost avoidance is not a hard savings, but cost removal is.

- *Soft savings* are not directly recognizable by accounting and/or the budget/P&L, but are reflected in indirect measurements, such as efficiency, customer satisfaction, time value of money, redirected labor effort, or savings in space, movement, and/or time, such as press/equipment time, cycle time, etc. Labor savings are soft savings unless they are factored in headcount reduction.

TEAM AND TIMELINE

Identify the roles, functions, skills, and resources needed. All major parties (stakeholders) affected by or affecting the project need representation. The project owner or champion must have control or assurance of the resources needed. If the assigned Belt doesn't have functional expertise in the problem area, subject matter experts (SMEs) need to be identified and acquired. Another thing that needs to be defined is a high-level project schedule, at least by phases.

Key Takeaways

- Effective project definition is crucial to ensure clarity, focus, and alignment to prevent misunderstandings and scope creep, setting the stage for success.
- A well-defined project requires a comprehensive project charter, which includes a clear problem statement, SMART objectives, a defined scope, stakeholder identification, a timeline, and resource allocation.

Chapter 4

Team and Change Management

With the project clearly defined, the next natural concern is to effectively manage the team and the changes the project will bring. Addressing the team and managing change at this point ensure the team is prepared for the journey ahead, which is essential for the project to succeed and for improvements to be sustained.

TEAM MANAGEMENT

TEAM means Together Everyone Achieves More. The typical team size for LSS is five to seven people. The team should be cross-functional. Consider roles and responsibilities, knowledge of the process, skill sets, diversity, and collaboration when selecting members. Outsiders can add unique value with fresh viewpoints.

Team Dynamics

A team is likely to go through four phases:

1. *Forming:* Team members are introduced and learn about the project and structure. They may feel excited, anxious, or curious. Team leaders can help by facilitating introductions and highlighting each person's skills and background.

2. *Storming:* Team members may experience obstacles and conflict as they push against boundaries and their true characters emerge. Leaders can help by welcoming and resolving conflicts and making time for healthy dialogue. The directive style of the leader can help the team move through this phase faster.

3. *Norming:* Team members learn to trust each other, ask for help, and resolve differences. They may appreciate each other's strengths and respect authority.

4. *Performing:* Team members are confident, excited, and satisfied with their work. They may be in a state of "flow" and performing at their full potential.

Critical factors for team success include:

- Purpose: Why do we have a team? What's our output?

- Process: What process will we follow?

- Communication: Is our communication complete and appropriate? What's the cadence?

- Commitment: How is commitment nourished?

- Involvement: Are the right people on the team?

- Trust: How is trust built and maintained?

Stakeholder analysis can significantly enhance the management of the team, project, changes, and communications. The interest-influence matrix can be used to identify the level of engagement with stakeholders—inform, consult, and collaborate—and establish a RACI chart. The interest-influence analysis can sort stakeholders into four buckets, as shown in Table 4.1.

Table 4.1   Interest-influence analysis.

Influence	Interest	Strategy
High	High	Key player; manage closely
High	Low	Keep satisfied
Low	High	Keep informed
Low	Low	Monitor (minimum effort)

RACI refers to four different roles:

1. Responsible: Someone who needs to complete the task.

2. Accountable: Someone who may not be doing the actual task but is ultimately answerable to others if the task is not completed.

3. Consulted: Those who help with getting the task done in collaboration with those responsible.

4. Informed: Parties who need to be kept up to date about the progress of a project, which may include receiving information about challenges and setbacks.

CHANGE MANAGEMENT

People in general don't like change. Some good options for dealing with resistance are identifying:

- A burning platform, or the reason for the change.
- "What's in it for me" (WIIFM), a tactic that identifies the personal-level connection or impact of a change to a person's self-interest.

There are various models for phases of change. One of the models involves *denial, resistance, exploration,* and *commitment.* They may not be sequential. People typically go through all four phases, likely at a different pace.

Practitioners have identified some key components for change to happen smoothly: dissatisfaction (reason for change), vision, plan, communication, etc. Precisely identifying, clearly documenting, and effectively communicating them can help get buy-in, reduce resistance, and sustain changes.

Key Takeaways

- The team and change management are critical for the successful execution of LSS projects.

- A team will likely go through four phases: forming, storming, norming, and performing.

- Change management addresses the impact of the change and helps manage resistance.

- Options for dealing with resistance include identifying a burning platform (reason for the change) and WIIFM.

Measure Phase

Once the improvement project is selected and clearly defined with the team in place, we progress to the Measure phase of DMAIC. When digging into a problem, the first thing you'll naturally do is get the relevant data, which is framed as Measure in Six Sigma.

It is worth noting that measure and analysis activities are typically mingled together, forming multiple iteration loops. Analyses typically prompt new data collection for deeper understanding. As explained earlier, we will cover non-statistical analysis activities in the Measure phase and statistical activities in the Analyze phase.

Chapter 5

Voice of the Customer and Data Collection

The first critical aspect and activity in the Measure phase is to understand the *voice of the customer* (VOC) and collect data relevant to the project. Although data collection happens throughout the project cycle, it's the focus of the Measure phase and thus is formally discussed in this chapter. This chapter will also discuss some popular techniques used to capture VOC and collect needed data to support identifying improvement opportunities and driving meaningful changes.

Who are the customers?

- *External* customers are those who receive the product, service, or output of the process and pay for the product or service.

- *Internal* customers are workers or others involved with the next step in the production process.

- *Stakeholders* include whoever is affecting or affected by the project.

WHAT IS VOC?

VOC is the customers' needs, expectations, preferences, and perceptions of a product or service that is delivered. It's a generic term and shouldn't be limited to customers only. Stakeholders' interests need to be considered as well; otherwise, they can cause the project to fail. Considering the scope of information needed for project success, VOC should cover all stakeholders; thus, it may be more accurately called the voice of the stakeholder (VOS).

Popular VOC and Data Collection Sources and Tools

Data collection should start by identifying what's needed for the project or problem and what's already available (the secondary/reactive data) to decide what to collect (the primary/proactive data) to bridge the information gaps. Primary data are the original data collected by a researcher for a specific purpose, while secondary data are data that have already been collected by someone else.

Secondary/reactive data include:

- Customer complaints, product returns, warranty claims
- Customer service calls, net promoter score, etc.

Primary/proactive data include:

- Surveys, which are systematic approaches to collect data. Surveys discover:
  - What's important to the customer
  - Customer satisfaction or dissatisfaction
  - Product or service quality, expected needs
  - Ways to improve product quality or services
- Interviews to learn the customers' views on product or service attributes and issues
- Focus groups, which offer qualitative research about people's perceptions, opinions, beliefs, and attitudes toward a product, service, concept, or advertisement
- Point-of-use observations, which see how a product or service is actually used. *Gemba* (現場) is a Japanese term that means "the real place" or "the actual place." A gemba walk (workplace walkthrough) can reveal additional insights not learned through other channels.

Key Takeaways

- Understanding VOC is crucial for aligning LSS projects with customer needs for project success.

- Data collection should start by identifying what's needed for the project or problem and what's already available (the secondary/reactive data), to decide what to collect (the primary/proactive data) to bridge the information gaps.

- Popular VOC tools include surveys, interviews, focus groups, and point-of-use observations.

Chapter 6
Diagnostic Process Mapping

Data are captured throughout the project, usually along with other activities. A core activity is to understand, visualize, and communicate the processes relevant to the project through *diagnostic process mapping* (DPM), which is primarily done in the Measure phase; thus, it's the next topic of discussion. DPM is essential for deep-diving into the problem to identify inefficiencies, bottlenecks, and areas for improvement. This chapter will explore popular mapping techniques and their roles in advancing LSS projects.

PROCESS MAPS

A process map is a visual diagram used to document, communicate, design, analyze, or manage a process. Why do we use process mapping?

- To study the problem through the lens of the process in map format. Determining which processes to study is usually quite obvious in manufacturing settings but not so in non-manufacturing settings, which requires some digging to identify what processes are relevant to the problem.

- To bring everyone involved to the same page with visibility to see the whole picture. Individuals usually see only a subset of the whole picture, and they see it differently from each other.

- To reveal associate-to-associate variation within the same activities.

- To enable everyone to highlight differences in the process from different angles and focus—for example, whether an activity is adding value from the customers' perspective.

There are typically at least three versions of a process:

1. What you think it is

2. What it actually is (current state)

3. What it should be (future state)

A process map is always a simplified version of reality. An effective process map in LSS reveals useful information to successfully advance our project. As the late statistician George Box said: "All models are wrong, but some are useful."

Popular Maps

All different mapping formats have pros and cons, each emphasizing different things. Which format to use is largely a personal preference. People tend to use what they are more familiar with to accomplish similar goals.

SIPOC Process Map

A SIPOC diagram is a mapping tool that visually documents a process with each process step identifying the Suppliers, Inputs, Outputs, and Customers, as shown in Table 6.1. SIPOC has an embedded cause-effect analysis in the IPO portion.

- SIPOC input
  - Process steps with a straightforward flow
  - Input information for each step and the suppliers' names
  - Outputs each step generates and the customer's name
- SIPOC output
  - Key process input/output variables (KPIVs and KPOVs)
  - Responsibility of suppliers and their capability to deliver inputs (availability), alignment to customer request
  - Gaps and improvement opportunities

Input variables can be classified based on their contribution to the process:

- C: *Controlled/Controllable* inputs, aka "knob" variables, which are input variables actively changeable to affect outputs

- U: *Uncontrolled/uncontrollable* inputs—for example, environmental variables such as humidity

  - Input variables that affect output variables but are difficult or impossible to control

  - Variables that are controllable yet are not being controlled actively

- K: *KPIV,* critical inputs, or input variables significantly impacting output variables (statistically proven)

Table 6.1 SIPOC example.

Supplier	Inputs	Process	Outputs	Customer
Outside sales	Information needed for purchase order (U, K)	Create purchase order	Purchase order	Inside sales
Customer service	Purchase order (C)	Create order summary	Order summary with all existing part numbers	Engineering
Engineering	Order summary, layout diagram, wiring diagram (C)	Generate new part numbers needed	Part numbers for all other line items	Manufacturing

Swim Lane Process Map

A *swim lane* is a visual diagram of process flow (flowchart) with an added dimension that arranges process steps in lanes by functions or responsibilities (that look like swim lanes) to make it easier to see who is responsible for certain steps (see Figure 6.1).

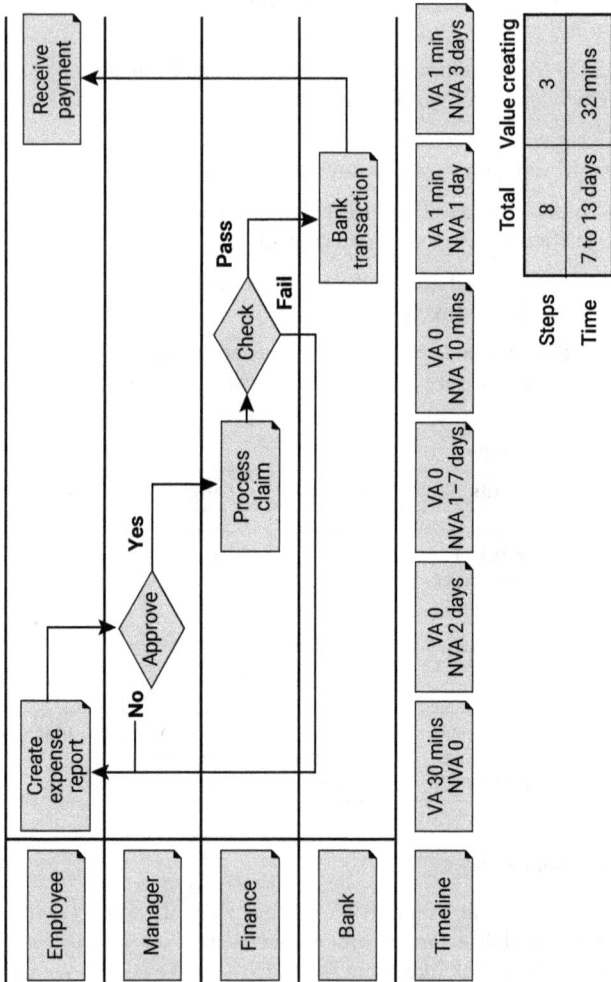

	Value creating	
	Total	
Steps	8	3
Time	7 to 13 days	32 mins

Timeline:

VA 30 mins NVA 0	VA 0 NVA 2 days	VA 0 NVA 1–7 days	VA 0 NVA 10 mins	VA 1 min NVA 1 day	VA 1 min NVA 3 days

Figure 6.1 Swim lane process map example.

- Swim lane input
    - Process steps, flow sequence, time needed for each step
    - Functions performing each step, each in a separate lane
    - Flowchart symbols that differentiate types of steps such as decisions and start or end
- Swim lane output
    - Throughput time (total time to complete the process)
    - Loops (rework), duplication, redundancy, or dead ends
    - VA (value-added) and NVA (non-value-added) steps, time, and variations
    - Gaps and improvement opportunities

Value Stream Map (VSM)

A value stream map (VSM) is a diagram of steps (with predefined data blocks, VA, and NVA) involved in the material and information flows to bring a product from order to delivery (see Figure 6.2). The current-state and future-state maps each depict the current or after-improvement's more ideal situation. One advantage of VSM is the format ensures end-to-end coverage of the process from supplier to customer, while some other formats may only cover a subset of the processes.

- VSM input
    - Product segmentation to distinguish the process
    - Customers, demand, and process steps in sequence
    - Material flow and handling information, information flow
    - Supplier information, inventory location, and level of material
    - Data block: process information such as cycle times, downtimes, capacity, availability, etc.

- VSM output
  - Customer takt time (the pace to make the product to meet customer demand), VA and NVA activities
  - Lead time, efficiency, and (in)balance of processes
  - Improvement opportunities ("starbursts" for kaizen) with waste and problem statements identified
  - Deployment plan to implement improvements

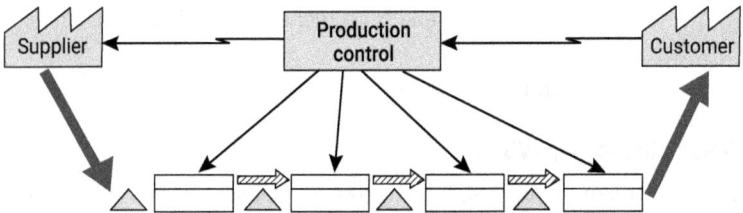

Figure 6.2 VSM example.

Spaghetti Diagram

A *spaghetti diagram* is used to illustrate the routes of information, materials, or people in a workspace (office, factory, or warehouse), so called because it looks like a messy plate of spaghetti (see Figure 6.3).

- Spaghetti diagram input
  - Layout or sketch of the location where the process takes place
  - Observation of the process flow
- Spaghetti diagram output
  - Traveled distance, time, and frequency of material, people, and/or information
  - Evaluation of the current-state layout of the workspace

– Number of steps, repetitive tasks, and flow reversal to complete the process

– Gaps and improvement opportunities

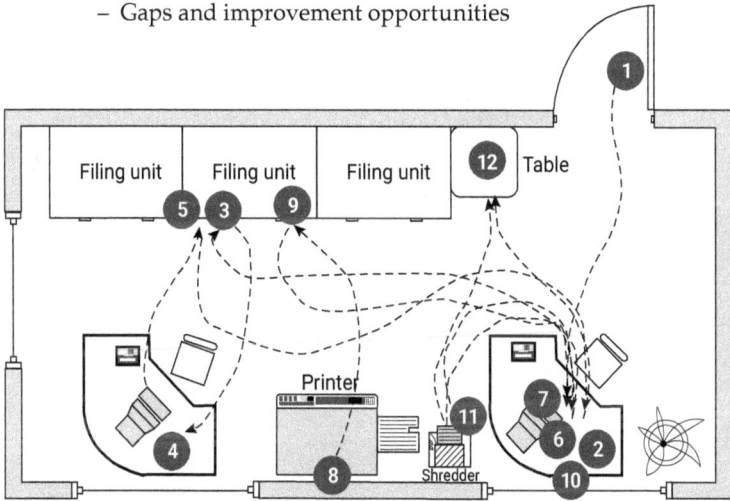

Figure 6.3   Spaghetti diagram example.

PROCESS MAP ANALYSIS

When reviewing process maps, look for:

- Differences between the documented process and the actual process

- Additional activities that need to be documented

- Differences between employees doing the same work

- Employees needing information or training

- Measurement points and measurements

- Cycle time of activities

- Problems, improvement ideas, and roadblocks

Key Takeaways

- Diagnostic process mapping is a core LSS activity, essential for visualizing processes and identifying inefficiencies and areas for improvement.

- Various mapping techniques, such as SIPOC, swim lane, VSM, and spaghetti diagrams, depict the process from different angles to look for opportunities.

- Effective diagnostic mapping involves defining objectives, gathering data, creating accurate maps, and engaging stakeholders.

Chapter 7
Root Cause Analysis

Having mapped out processes and identified areas for improvement, the next step, also a core activity in LSS, is to delve deeper to understand the underlying root causes (RCs) of the issues and inefficiencies. Cause-effect (C-E) analysis, also known as *root cause analysis* (RCA), is an essential tool in Measure and Analyze phase iterations. Digging deeper into causes may prompt new data collection needs. This chapter explores various techniques for identifying RCs, to ensure solutions address the RCs rather than the symptoms.

THE TRUE MEANING OF RC

The concept of an infinite chain of causation is that the cause-effect relationship relies on the law of nature and physics to establish, forming an infinite chain with no end. There is always a cause to any event. The same item is both a cause of its downstream and an effect of its upstream at the same time. There is no "true" absolute RC per se. A so-called "root cause" is something subjectively chosen to serve that role. An RC is what is chosen to be, not born to be. There is no entitlement.

THE POINT OF DOING RCA: A PARADIGM SHIFT

The traditional RCA theme urges people to continuously migrate the focus to the upstream causes, which is the concept of the so-called 5 *Whys*. Why continuously shift the focus to the upstream causes? Because the impact of upstream causes is bigger in general. Yet, there is an unwanted companion for going upstream: the research cost will also increase along the way, which is a challenge! With an infinite chain of C-E, how do we know where to stop and claim RC? This is a constant struggle and dispute for many people.

Let's take a look at the Jefferson Memorial Case to feel the pain. The U.S. Park Service noticed the Jefferson Memorial monument was deteriorating faster than the other monuments. They investigated it using the 5 Whys and formed a chain of causation:

Why? Because it's washed more frequently.

Why? Because it has more bird droppings.

Why? Because more birds are attracted there.

Why? Because there are more fat spiders.

Why? Because more midges fly there in the evening.

Why? Because the illumination attracts midges.

The solution chosen by the park service: turn on the lighting one hour later in the evening. This simple act allegedly reduced 90 percent of the bird-dropping problem. For a more detailed discussion, refer to Jing, G. (2008).

Three questions can be asked to answer in this case:

1. Is the "true" root cause found?

2. Is the solution (the selected RC) the only choice?

3. Is the solution a good choice?

The 5 Whys can keep going on and on. Multiple events at various stages can be declared as RC; thus, various solutions can be sought to break the chain of causation. Yet, the chosen solution is a good one because of its low cost and high return.

So, where do you stop (to cut to the chase)? The trick is not to find the "true" root cause; rather, it is to find the leverage point that benefits the problem the most and then call it the RC. The key is ROI—low cost with high return. This is what people do subconsciously without realizing it. ROI is missing in most RCA doctrines, which is a major deficiency of most RCA practices—but it has always been at play, mostly unnoticed.

ROI LEVERAGE POINT

A lot can be said about ROI assessment. In a way, RCA is about ROI calculation, intentionally or unintentionally. Whether they realize it or not, that's what goes through people's minds when choosing RC. The trick is to find an RC with a solution that has a significantly lower implementation cost to offset the increasing research cost, which lowers the overall solution cost, as shown below:

Solution Cost ↓ = Research Cost ↑ + Implementation Cost ↓↓

ROI assessment could be short term or long term, internally (self) or externally (customer) oriented. The results could be very different. Short-term fixes are mostly driven by cost constraints. ROI may change over time, and so does the choice of RC. To be long lasting or sustainable, all parties' (stakeholders') interests need to be balanced.

Here are three tips to find the leverage point:

1. Follow the *Pareto principle,* often called the 80/20 rule: 80 percent of the problems are caused by only about 20 percent of the contributing factors.

2. Use a desirability matrix for finding the leverage point (see Chapter 2).

3. The span-of-control/sphere-of-influence can come in handy (see Chapter 2). RCs are ideally within the span of control or at worst within the sphere of influence.

SCOPE OF THE PURSUIT VS. SELF-IMPOSED CONSTRAINT

Staying within the scope of the pursuit improves ROI and leverage. Going beyond the intended scope causes distractions, dilutes the effort, and diminishes the return. Yet, the self-imposed scope also serves as a constraint, which may prevent longer-term solutions.

POPULAR RCA TECHNIQUES

Here are five techniques, sorted from simple to complex, which should cover most of the RCA needs:

- Is/is-not comparative analysis to zoom in through segregation
- 5 Whys to go deep

- Fishbone diagram (C-E diagram), a traditional approach for brainstorming and diagraming C-E relationships, which is good for studying one primary effect—to go broad

- Root cause tree (RCT), a problem analysis diagram that combines fishbone and 5 Whys, allowing the study of the relationships between causes—to go complex

- Cause-effect (X-Y) matrix, used to study the relationship of a group of causes (inputs, X) and a group of effects (outputs, Y) and quantify the relationships/impacts

Is/Is-Not Comparative Analysis

As said by Rudyard Kipling: "I keep six honest serving men. They taught me all I knew. Their names are What, Why, How, Where, When, and Who." The is/is-not analysis (see Table 7.1) isn't a traditional RCA tool, but it can serve the need through divide and conquer, especially during the early stages of projects when defining the problem.

Table 7.1   Is/is-not comparative analysis.

Problem statement:

	Is	**Is not**	**Differences and changes**
What	What is the specific object that has the defect? What is the specific defect?	What similar objects could have the defect but do not? What other defects could be observed but are not?	
Where	Geographically? Physically on the part?	Where, when, and what size could the defect have been, but it was not?	
When	When was the defect first observed? When since then? When in the product life cycle?		
Size	How many objects with the defect? How many defects per object? Size of defect? Trend?		

5 Whys

See the Jefferson Memorial example on p. 40 for the simplest RCA technique.

Fishbone Diagram

Also called the Ishikawa diagram, a fishbone diagram is one of the most popular RCA tools. It focuses on a single effect with simple C-E relationships and displays major categories in the shape of a fishbone, which is suited to search broadly but not deeply (see Figure 7.1).

- Advantages: Helps organize factors, provides a structure for brainstorming, and involves everyone.

- Drawbacks: Might become very complex, and does not rank the causes in an if-then manner nor show relationships between causes.

Commonly used fishbone categories:

- For product: 5Ms + E. Man, Machine, Method, Material, Measurement + Mother Nature (Environment)

- For transactional process: 4Ps + M&E. People, Policies, Procedures, Place, Measurement+ Mother Nature (Environment)

One simple way to improve the effectiveness of the fishbone diagram is to add some quick ROI assessments to each identified cause right on the diagram using a two-letter notation (R, C):

- Return/opportunity (R): H/M/L for high/medium/low.

- Cost/uncontrollability (C): H/M/L for high/medium/low (H, L) indicates a favorable high ROI, while (L, H) indicates an unfavorable low ROI.

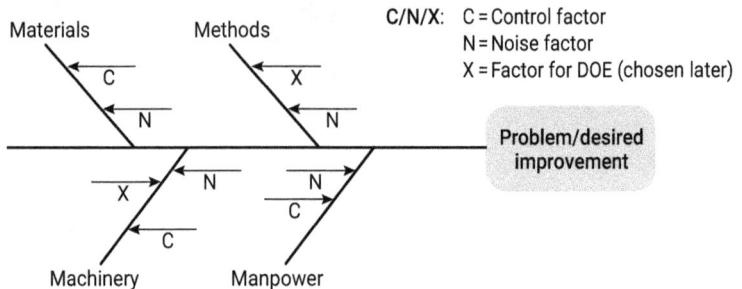

Figure 7.1 Fishbone diagram.

Root Cause Tree (RCT)

RCT combines fishbone and 5 Whys for more complicated C-E relationships (for example, causes may be dependent on each other), enabling the team to go both broad and deep (see Figure 7.2).

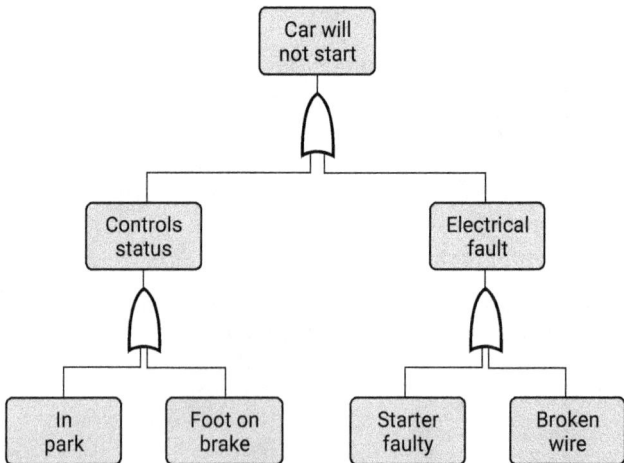

Figure 7.2 RCT example.

How to generate the tree:

- A *top-down* approach (more structured): Use a fishbone diagram first, and if it's not enough, turn the bone structure 90 degrees counterclockwise to form the initial tree; study the (and, or) relationship among the existing causes, and then expand from there to the next tier. Mind mapping can be used instead of a fishbone to generate ideas in a more flexible format.

- A *bottom-up* approach (less structured): Use an affinity diagram to generate and organize ideas, and then document in a tree structure. An affinity diagram is a technique to organize various ideas into their natural relationships.

Cause and Effect Matrix

The C-E matrix, aka X-Y matrix, evaluates and quantifies the impact of a group of causes (inputs, X) on a group of effects (outputs, Y). It uses the weighted summation of the individual impact of each cause (input variable) against each effect (output variable) to get the overall impact of each cause against the whole group of effects. The overall impact can be used for prioritization. In Table 7.2, the total impact of customer input is the summation of the multiplication of the individual impact of customer input and the importance of outputs/effects:

$$8 \times 10 + 6 \times 6 + 8 \times 9 + 8 \times 8 = 252.$$

Table 7.2   C-E matrix example.

	Importance	**10**	**6**	**9**	**8**	**Total**	**%**
	Outputs/effects	1	2	3	4		
	Inputs/causes	Efficiency	Commonality	Yield	Implementation		
1	Customer input	8	6	8	8	252	39%
2	Equipment specs	5	5	10	8	234	36%
3	Bill of materials	7	5	4	3	160	25%
						646	

Key Takeaways

- There is no truth of RC per se. An RC is subjectively chosen to serve that role.

- The point of doing RCA is to look for the leverage point for high ROI.

- Popular RCA tools, from simple to complex, include is/is-not, 5 Whys, fishbone, RCT, and C-E matrix.

- Effective RCA involves defining the problem, brainstorming the potential causes, identifying the root causes, verifying the root causes, and developing solutions.

- Solutions based on RC with high ROI are more likely to be effective and sustainable, leading to lasting improvements.

Chapter 8

Failure Mode and Effects Analysis (FMEA)

After conducting RCA, a natural progression is to start a *failure mode and effects analysis* (FMEA), another core activity in LSS projects, although the final iteration of FMEA will finish in the Control phase. FMEA is a systematic method for identifying and prioritizing potential failure modes in a process, product, or system and their effects on performance, as well as the current controls and possible mitigations. While RCA helps us understand the RCs of problems, FMEA expands the understanding by anticipating and preventing potential problems from occurring. One advantage of starting FMEA at this stage is to have a good visibility and understanding of current controls (and potential mitigations). The mechanism of FMEA is rather simple; it can be done by filling in the FMEA form. Yet to do it well isn't easy, so let's dive in.

TYPICAL STEPS IN FMEA

- *Define the scope:* Clearly define the scope of the FMEA, including the process, product, or system being analyzed.

- *Identify the failure modes* (FM): Brainstorm the potential failure modes that could occur in the process, product, or system.

- *Assess severity* (S = the seriousness of the effect), *occurrence* (O = the likelihood of the failure occurring), and *detection* (D = the probability of the failure being identified before causing an effect): Rate the S, O, and D of each failure mode on a scale of 1 to 10. Note that for detection, easy detection gets a low rating, which may be overlooked by new users.

- *Calculate RPN:* Calculate the risk priority number (RPN) for each failure mode (RPN = S x O x D).

- *Prioritize actions:* Prioritize actions based on the RPN, focusing on high-risk failure modes first.

- *Develop action plans:* Develop action plans to mitigate or eliminate high-risk failure modes, assigning responsibilities and timelines.

Different industries have different expectations of FMEA. Highly regulated industries, such as medical device manufacturing, aerospace, or automotive, are more serious and formal with FMEA, which requires them to follow their prescribed format. Unregulated industries or SS activities are more flexible and are open to customization or less formal formats, such as potential problem analysis (PPA) or risk register (see FMEA-lite example, p. 50).

WHAT ARE THE BIGGEST CHALLENGES IN PERFORMING FMEA?

Effectiveness and efficiency are a concern; the quality of FMEA is hard to assess at the time of doing it since effectiveness can only be verified in the future. With huge investments and no guarantee of results, people tend to cut corners.

There is a fundamental deficiency and limitation in traditional FMEA and RCA: the cost of actions is usually not considered, while cost can never be ignored in business decisions. Such a fundamental gap is unrecognized by most people, which has created inherent problems and struggles in many FMEAs and RCAs, making them less effective.

Which Parts of FMEA Are Difficult?

- Identify the type of FM, which may include:
  - Failing to meet the specification
  - Incorrect or inappropriate requirements
  - Unintended use, application, or environment
  - Doing harm to others

- Is it a cause or an effect? It depends. The C-E role will shift, depending on the level in the chain of causation. For example, an electrical fault can be a cause at the car level, a failure mode at the subassembly level, and an effect at the component level.

- Identify the type of control: Is it prevention or detection?

  - Prevention focuses on preventing a failure from occurring, affecting occurrence (O).

  - Detection focuses on preventing the effect from occurring, affecting detection (D).

- Is it a process issue or a design issue? Some failure modes may belong to both. Many process issues can be better addressed via design changes.

  - DFMEA focuses on potential failures in product design.

  - PFMEA focuses on potential failures in the manufacturing process.

- Interaction between (sub-) structures (IBS) are hard to capture, especially in FMEA format. Yet, studies show approximately 70 percent of failures are related to interfaces.

Tips on FMEA Efficiency and Effectiveness

FMEA components aren't equally important. Which ones are the most important?

- The most important component is FM (concerns). Identify potential failures as thoroughly and completely as possible.

- The second most important component is the countermeasures (actions) to contain the risk of FMs at an acceptable level. Be conscious of the cost, though.

- The third most important is the RPN to assess the risk level and drive actions. Usually, a threshold (for example, 125) is set for RPN, which requires action, if exceeded. (For example, 125 = all risk components at middle level 5.) Yet, RPN alone isn't sufficient in directing actions. Some practices also require action when the severity is high, such as 10. In the meantime, ROI is another driver. Part of the reason is that RPN is always highly subjective due to data constraints. The RPN only adds value when the risk level is in dispute. For undisputed situations, a simple assessment of risk level should be sufficient to drive actions:

- High = actions required

- Low = actions not required

- Medium = actions optional/discretionary

- Everything else in FMEA is in the supporting role to help make informed decisions of the above, only realizing (or adding) value when leading to different actions (or FMs).

Redirect the team's brain power to boost FMEA efficiency and effectiveness:

- Focus team resources on the top two most important components (FM and countermeasures) that can benefit the team the most. An experienced lead person or a small sub-team can handle the rest well.

- Focus on NUD (new, unique, difficult) and ease on ECO (easy, common, old) items. The key contents of FMEA may be identified outside of FMEA forms, which makes FMEA just a documentation and communication format in this case.

FMEA-LITE

FMEA-lite is a tiered triage option to deal with a lack of data in a high-mix-low-volume environment when FMEA is not used for extreme risk aversion and is not regulated by agencies or customers. This practice may be controversial, but it surely is thought-provoking.

The FMEA-lite practice is defined as follows: When the condition doesn't support deep-level risk assessment, as mentioned previously, this tiered triage approach uses a simpler risk analysis technique to do an initial risk assessment to improve efficiency and effectiveness, entering results in an FMEA form without necessarily utilizing all fields on the form and switching back to traditional FMEA practice for post-action risk assessment.

Lack of data is a major amplifier to inefficiency, typically seen in high-mix-low-volume environments or early stages of product development. A 10-level resolution may be unnecessary or overkill in this case since it introduces more noise and convolutes signals. In *The Paradox of Choice—Why More is Less* (Schwartz, 2005), the author explains the psychology behind how eliminating consumer choices can greatly reduce anxiety for shoppers. To deal with the lack of data, one can use three-level scoring. Another option is to go further to simpler risk analysis.

The simpler risk analysis (see Table 8.1), such as risk register, typically involves fewer risk components and fewer assessment levels. Generally you'll see three (high-medium-low or H-M-L) or five levels, which is widely used in project management. A more scaled-down version may use a single risk rating. It's much easier to sort things into three buckets than 10, especially when lacking data.

Table 8.1   Example of simpler risk matrix.

		Consequence/impact		
		Slightly harmful	**Harmful**	**Extremely harmful**
Likelihood	Likely	Medium risk	High risk	Extreme risk
	Unlikely	Low risk	Medium risk	High risk
	Highly unlikely	Insignificant risk	Low risk	Medium risk

FMEA-lite Steps

1. Brainstorm risk items. Discuss/clarify the current state and gap for each item. Focus on NUD (new, unique, difficult) items.

2. Quickly assess risk level (H-M-L).

   a. If in easy consensus, it may not need to be broken down into risk components.

   b. If in question or dispute, assess the likelihood and impact with H-M-L. Optionally, consider the current control level (detection), since the *impact* can be considered as the combination of *severity* and *detection*.

   c. If there is no information at all about a risk or component, M should be the default level.

3. Decide if the team wants to do anything per the risk.

   a. General rule: H = action required; M = action optional/ discretionary; L = action not required.

   b. To be more aggressive, the M level can require action.

   c. The team may choose to take action on some low-risk items when the cost is low and ROI is high—the second driver for actions. You may also have low confidence in risk assessment rooted in data deficiency. Once this occurs, a low-risk item may no longer be low.

4. A lead person can enter contents into the FMEA form.

   a. Risk item go in the FM field. You may enter only items requiring actions to help efficiency.

   b. Risk rating goes in the RPN-related fields. You can use the following simple conversion as an example:

      i. If not, break down RPN: H = 300, M = 125, L = 50.

      ii. If to, break down RPN, for impact (severity) and likelihood (occurrence): H – 8, M – 5, L – 3. Use 5 for detection for simplification.

   c. Action item to action field.

   d. Populate the remaining fields as needed (or if required or requested).

Key Takeaways

- FMEA involves defining the scope; identifying failure modes; assessing severity, occurrence, and detection; calculating the RPN; prioritizing actions; and developing action plans.

- Efficiency and effectiveness are the two biggest challenges in FMEA. Focusing on the two most important components—FM and actions—can help efficiency.

- When lacking data, a simpler risk analysis or the FMEA-lite version can be a better choice to assess risk.

Analyze Phase

Along with data collection, a natural activity is to analyze the data, which is framed as the Analyze phase in SS. As mentioned earlier, there are likely many Measure-Analyze iteration loops in a project, as analyses tend to prompt new data collections. In general, findings from non-statistical analyses should be validated or verified by data, and analytics may prompt a new need for data. Therefore, many analytical tools are used in both phases. In this book, statistical data analytical tools are found in the Analyze sections.

Nowadays, data analytics typically relies on widely existing statistical software for such capabilities. This book will only touch data analytics lightly, focusing on the unique analyses used by SS and lightly touching on several generic statistical analyses. It will highlight what types of data analytics are typically useful for LSS projects but will not cover how to complete them using software. *NIST/SEMATECH Engineering Statistics Handbook* (https://www.itl.nist.gov/div898/handbook/) is a free resource for more in-depth detail and examples of applications.

Graphical analysis (data visualization) is the first data analysis that all LSS projects need to run. In addition, there are three signature SS (or quality) data analyses that are not typically covered in generic statistical analyses: SPC, MSA, and capability analysis. This section will also lightly touch on hypothesis testing and regression and will end with a simple DOE introduction.

Chapter 9
Data Visualization

The very first data analytics should be *data visualizations* to uncover patterns and trends that may not be apparent in the raw data. By visualizing data, we can gain deeper insights into the data and process, and decide on the next meaningful analytics.

DATA MINING AND GRAPHICAL ANALYSIS

Data mining refers to finding suitable data to explain why a product or process behaves in a certain way. It follows some basic steps:

- Look at the raw data to identify any abnormalities (errors, unexpected values, etc.). We can call this a *sanity check*.

- Analyze the data graphically to look for phenomena that stand out.

- Analyze the data statistically to confirm or reject the observed phenomena.

There are four dimensions to study the data visually (see Table 9.1): distributions, trends, relationships, and priorities. Which analysis to use is situational, depending on the need.

Table 9.1   Four types of graphical analyses and tools.

Distributions	Trends	Relationships	Priorities
• Dot plot • Histogram • Box plot	• Time series plot • Run chart • Individuals chart	• Scatter plot • Graphs with groups	• Pareto chart

DISTRIBUTIONS

Distribution can visualize the three most important statistics of data: center, spread, and shape. Note that *shape* is an important piece of information, since most statistical analyses need to know if the data are normal or not.

Dot Plot

In a dot plot (see Figure 9.1), each dot represents an "event" of output at a given value on the X-axis. As the dots accumulate on the Y-axis, the shape of the dots represents the distribution of the data.

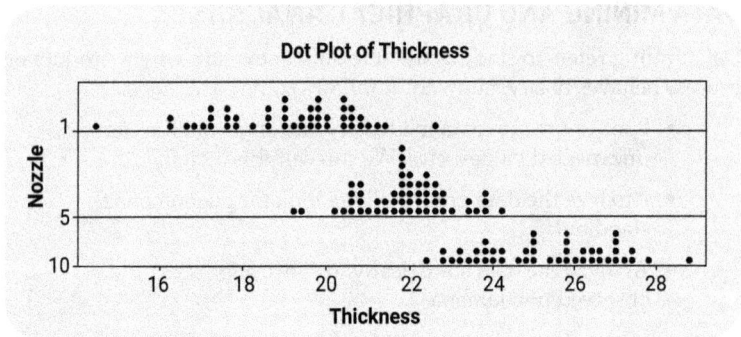

Figure 9.1 Dot plot.

Histogram

A histogram (see Figure 9.2) groups data into intervals, each with a bar representing the associated data frequency.

Box Plot

A box plot (see Figure 9.3) shows the spread and center of the data in a more abstract way than the dot plot. Note that the center line of the box plot is the *median*, not the *mean*. The first and third quartiles (Q1 and Q3) correspond to the 25th and 75th percentiles of the data from the minimum, respectively.

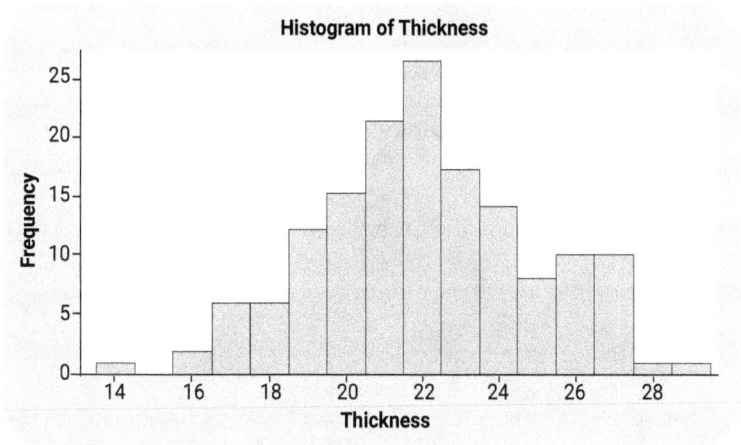

Figure 9.2 Histogram.

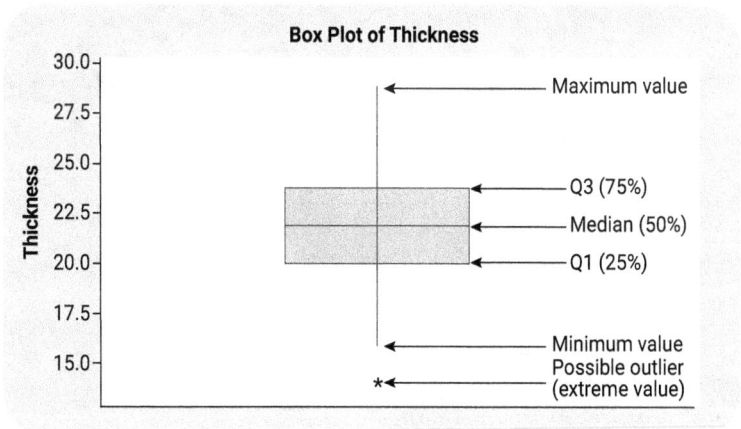

Figure 9.3 Box plot.

Trends

A time series plot plots each data point along a timeline to show the trend and profile of data over time. A run chart (see Figure 9.4) adds a standard set of pattern analyses to the time series plot, including a statistical assessment of the following (significant when p-value < .05):

- Clusters: A group of points that are close together in one area of the chart, indicating a shift in the mean. Clusters may be caused by special causes like measurement problems, sampling, or defective parts.

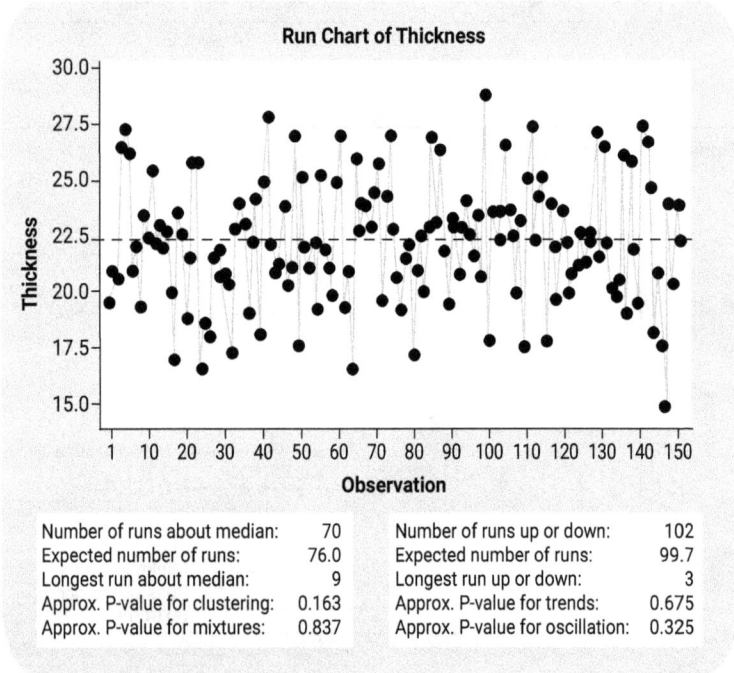

Number of runs about median:	70	Number of runs up or down:	102
Expected number of runs:	76.0	Expected number of runs:	99.7
Longest run about median:	9	Longest run up or down:	3
Approx. P-value for clustering:	0.163	Approx. P-value for trends:	0.675
Approx. P-value for mixtures:	0.837	Approx. P-value for oscillation:	0.325

Figure 9.4 Run chart.

- Mixtures: A pattern of frequent crossing of the center line, indicating a lack of data points near the center line. Mixtures may be caused by a bimodal distribution due to changes in shift, machinery, or raw materials.

- Trends: A sustained drift in the data, either up or down. Trends may be caused by special causes like tool wear, a machine not maintaining configuration values, or operator rotation.

- Oscillations: Rapid up-and-down fluctuations in the data, indicating process instability. Oscillations may look like a sine wave.

The individuals chart (I-chart), a popular SPC chart, adds control limits to the time series plot. The I-chart is presented in more detail in Chapter 10.

RELATIONSHIPS/COMPARISON

Certain tools are used to study comparisons within data points.

Scatter Plot

Scatter plots (see Figure 9.5) are used to study the correlation between two variables. Many software programs can provide R^2, the coefficient of determination, which measures the strength of the correlation. The range of R^2 is between 0 and 1, with higher values indicating a stronger correlation. A more detailed discussion of various correlations is presented in Chapter 14.

Graphs with Groups

Graphs with groups provide a visual comparison of the three most important statistics of data: center, spread, and shape (see Figure 9.6).

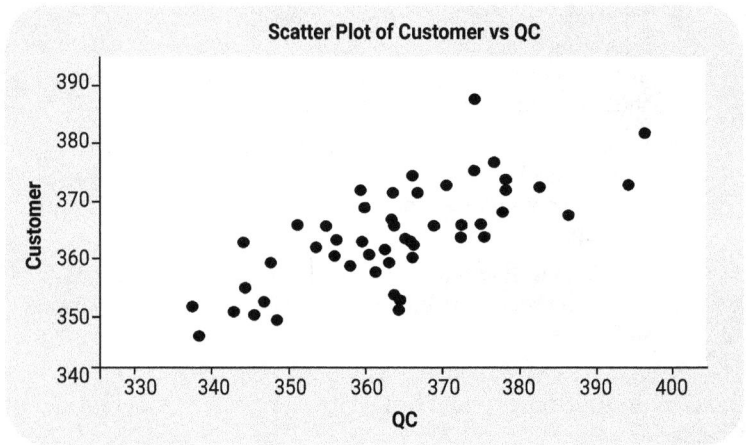

Figure 9.5 Scatter plot.

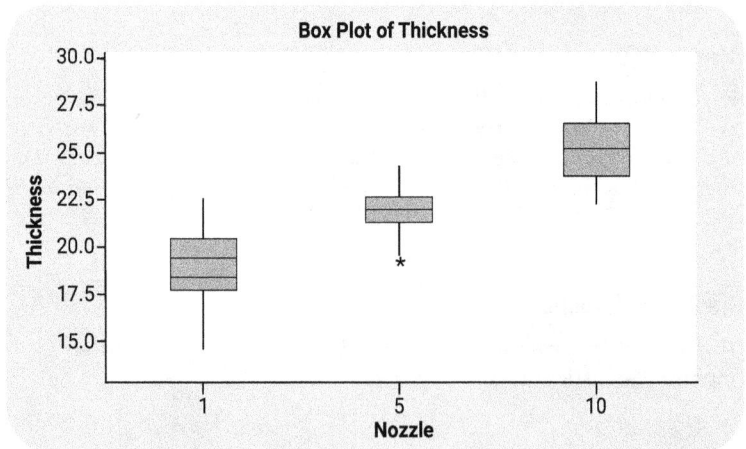

Figure 9.6 Box plot with groups.

Priorities (Pareto)

The Pareto chart (see Figure 9.7) sorts categories by frequency and is an essential tool for prioritizing improvement opportunities. The 80/20 rule states: 20 percent of the causes are responsible for 80 percent of the problems.

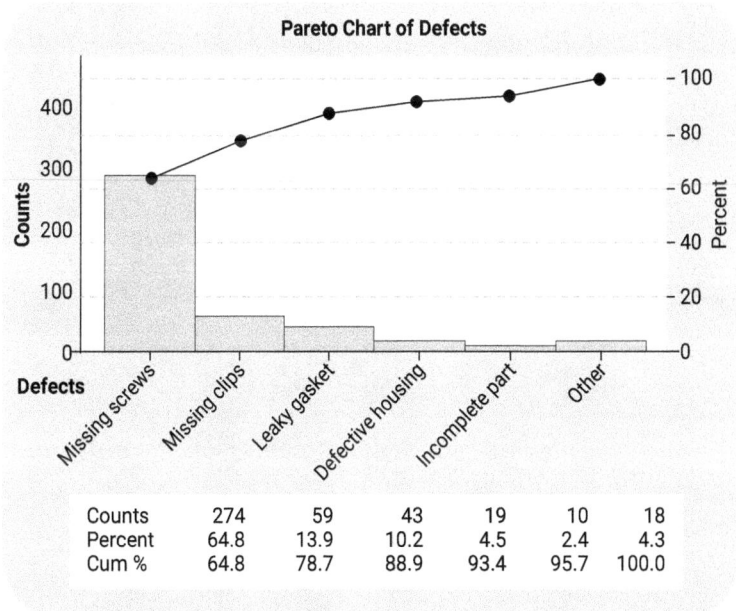

Pareto Chart of Defects

	Missing screws	Missing clips	Leaky gasket	Defective housing	Incomplete part	Other
Counts	274	59	43	19	10	18
Percent	64.8	13.9	10.2	4.5	2.4	4.3
Cum %	64.8	78.7	88.9	93.4	95.7	100.0

Figure 9.7 Pareto chart.

Key Takeaways

- Data mining involves looking for phenomena graphically and validating the phenomena statistically.

- Data visualization helps identify patterns, trends, and outliers in data, leading to informed decision-making.

- There are four dimensions to visually study the data: distributions, trends, relationships, and priorities.

- Distribution visualizes the three most important statistics of data: center, spread, and shape. Thus, it is usually the first step in analysis.

- By incorporating data visualization into an LSS project, we can gain deeper insights into process performance and identify improvement opportunities more effectively.

Chapter 10
Statistical Process Control (SPC)

After getting some idea about the data visually, a natural next step is to gain a deeper understanding of how stable the data (or process) are over time. This chapter will discuss *statistical process control* (SPC), or control charts (CC), to monitor and control process variation and stability, which is an extension of trend analysis.

TWO TYPES OF VARIATION

- *Random or common cause (noise)* variation is produced by the process itself (the way things naturally occur), present in every process. Reduction or removal requires fundamental changes in the process. A process is considered in control, stable, and predictable if it is only exhibiting noise.

- *Assignable or special cause (signal)* variation is caused by unique disturbances that are unpredictable and typically larger than noises. It can be removed or lessened via typical process control and monitoring activities. A process is considered out of control and unstable if it exhibits special cause variation.

STATISTICAL BASIS FOR SPC

The general model for the center line and upper/lower control limits (UCL/LCL) for variables charts is

$$\text{UCL} = \overline{X} + k\overline{s}$$

$$\text{Center line} = \overline{X}$$

$$\text{LCL} = \overline{X} - k\overline{s}$$

where

- $\overline{X}$ = Sample mean, $\overline{s}$ = Sample standard deviation
- k = Number of $\overline{s}$ that the control limit is away from the center line (usually 3)

When only common cause variation is present in the process, 99.7 percent of the data should be within the control limits. If any data points are outside the control limits, it's usually an indication that special cause variation is present.

TWO BIG CONTROL CHART MISTAKES

- Putting specification limits on a CC
- Treating UCL and LCL as specification limits (USL/LSL)

When either case happens, the CC becomes an inspection tool and is no longer a CC. UCL/LCL are not directly tied to USL/LSL. Process control limits are calculated based on data from the process itself, while specification limits are provided by the customer. Deciding how a process performs against customer requirements is a capability analysis. Ideally, control limits should fall inside the specification limits to avoid making defects (see Figure 10.1).

POPULAR TYPES OF CONTROL CHARTS

- The *X-bar chart* uses the mean of a subgroup of variable data, and the R chart uses the range of the subgroups to show changes over time.
- An *individuals chart* (I-MR) monitors the variation in individual data points, where MR refers to the moving range (the distance between two consecutive points).
- *P charts* monitor the proportion of defective items in a process.
- *C charts* monitor the number of defects in a consistent sample size.

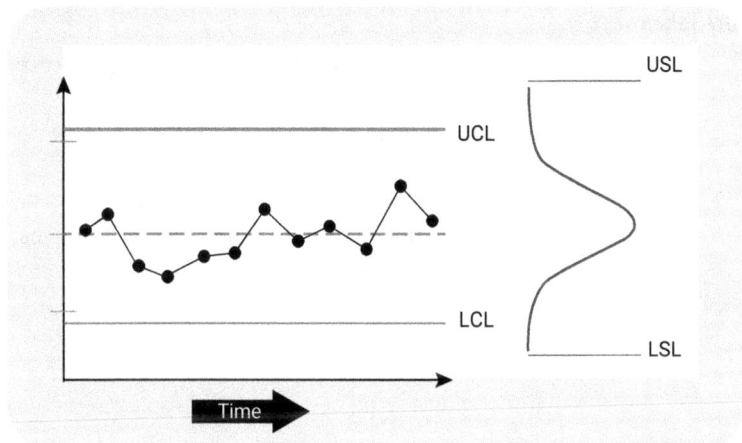

Figure 10.1 UCL and LCL vs. USL and LSL.

CONTROL CHART RULES

Below is a set of rules to identify special cause events. Most statistical software has them embedded in analysis; thus, there is no need to apply these rules personally. When a rule is broken, the process is "out of control;" that is, something unusual has happened, which needs to be investigated!

1. One point is outside the UCL or LCL (three-sigma limit).

2. Two of three consecutive points are outside the two-sigma limit.

3. Four of five consecutive points are outside the one-sigma limit.

4. Eight consecutive points are on one side of the center line.

5. A trend emerges; six consecutive points are either increasing or decreasing.

6. A pattern occurs; a cycle appears, or a pattern repeats itself.

Key Takeaways

- SPC is a powerful tool for monitoring and controlling process variation and stability.

- Popular control charts include the X-bar and R chart, I-MR chart, P chart, and C chart.

- If a process isn't stable, action needs to be taken to stabilize the process before proceeding to the next stage for improvements.

- The stabilizing process itself is an improvement.

Chapter 11

Measurement System Analysis (MSA)

After getting a good understanding of the data and stability, next we'd like to know how much we can trust the data. This brings us to the *measurement system analysis* (MSA), or gage R&R (GR&R), which helps us assess the reliability and validity of our measurement systems and data.

WHAT IS MSA?

Measurement system analysis assesses the accuracy, precision, linearity, and stability of the measurement system. Among these, precision is usually the biggest challenge to deal with. It is primarily measured by gage R&R; thus, MSA is also referred to as GR&R.

ACCURACY VS. PRECISION

- *Accuracy* is how close the measured values are to the standard that was measured. It's assessed through calibration against a given standard. Bias is the difference between a known value (standard) and the measurements obtained from an instrument.

- *Precision* is the variation within the repeated sets of measurements of the same feature.

Repeatability and Reproducibility

Precisions have two types:

1. Repeatability σ^2_{rpt}, or gage variation, is the variation in repeated measurements when nothing has changed.

2. Reproducibility σ^2_{rpd} is the variation in repeated measurements when something has changed (most typically the operator). Other changes may include equipment, method, location, environment, or time. The measurement feature cannot be changed.

Components of Variation

The components of variation are determined through a gage R&R study:

$$\sigma^2_{total} = \sigma^2_{part} + \sigma^2_{ms}$$
$$\sigma^2_{ms} = \sigma^2_{rpt} + \sigma^2_{rpd}$$
$$\sigma^2_{rpd} = \sigma^2_{op} + \sigma^2_{opxpart}$$

where σ^2_{total} is total variation, σ^2_{part} is part-to-part variation, σ^2_{ms} is measurement system variation, σ^2_{rpt} is repeatability, σ^2_{rpd} is reproducibility, σ^2_{op} is operator-to-operator variation, and $\sigma^2_{opxpart}$ is operator x part interaction.

Measurement System Performance Metrics

- Metrics of accuracy or bias
- Metrics of precision, measured from various angles:
  - *Precision-to-tolerance ratio* (P/T, percent tolerance) is the fraction of the tolerance consumed by measurement system variation, an external-looking metric against specifications.
  - Gage repeatability and reproducibility (GR&R) is the fraction of total variation consumed by measurement system variation. It is an internal-looking metric that can be measured in two ways:

1. By variance, referred to as percent R&R, percent contribution, or "% VarComp" in Minitab

2. By standard deviation, referred to as P/TV, percent process, or "% Study Var" in Minitab

– The number of distinct categories or the number of categories the measurements can be grouped into, an internal-looking metric equivalent to GR&R.

Both internal- and external-looking metrics need to be good for a gage to be good. Otherwise, some aspects of the gage aren't good. Below are some evaluation guidelines:

- For P/T, percent tolerance, or percent process ("% Study Var" in Minitab):

  – Less than 10 percent = good

  – Between 10 percent and 30 percent = acceptable, subject to the application, the cost of the measuring device, the cost of repair, or other factors

  – Greater than 30 percent = unacceptable; improvement needed

- For percent R&R (percent contribution, "% VarComp" in Minitab):

  – Less than 1 percent = good

  – Between 1 percent and 9 percent = acceptable, subject to the application, the cost of the measuring device, the cost of repair, or other factors

  – Greater than 9 percent = unacceptable; improvement needed

- For the number of distinct categories: ≥ 5 is good.

Key Takeaways

- MSA is essential for assessing the reliability and validity of measurement systems and data, leading to better decision-making.

- Common MSA studies include gage R&R, bias, and linearity and stability studies.

- If gage R&R isn't good, improvements need to be made to the measurement systems before proceeding to the next stage for improvements.

- MSA improvement itself is an improvement for the project.

Chapter 12
Capability Analysis

After validating the measurement systems and verifying that the process is stable, a natural next question is how well we are performing against the specifications or customer requirements. This brings us to the current chapter. *Capability analysis* assesses the ability of our processes to meet customer requirements, represented by USL and LSL, the upper and lower specification limits. This can only be done reliably after assuring good measurement systems and process stability.

C_p, C_{pk}

C_p is referred to as the "entitlement" performance, the best the process can perform if the process is centered on the target. C_p can't tell the practitioners whether the process is centered on the specification target t. The commonly adopted expectation is $C_p \geq 1.3$.

$$C_p = \frac{USL - LSL}{6\sigma}$$

C_{pk} reflects the actual performance of the process by taking the worst performance between the upper and lower sides. The difference between C_p and C_{pk} indicates whether the process is "off-target" (see Figure 12.1): it is on-target if $C_p = C_{pk}$ and off-target if $C_p > C_{pk}$. C_p can never be lower than C_{pk}.

$$C_{pk} = \text{Min} \left(\frac{\overline{X} - LSL}{3\sigma}, \frac{LSL - \overline{X}}{3\sigma} \right)$$

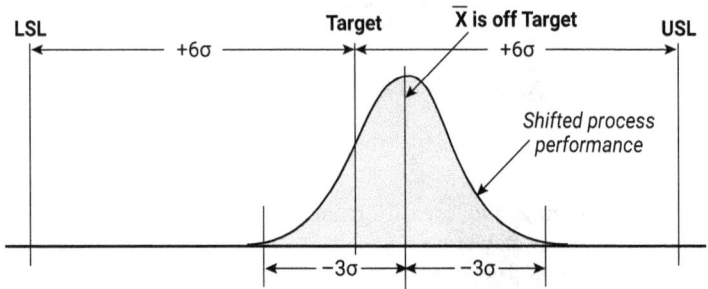

Figure 12.1 Off-target process.

Z STATISTIC

Z represents the number of standard deviations between the distribution average and the specification limits of a normally distributed process output, often referred to as the "sigma level" of performance. The 1, 2, and 3 sigma level (range) of normal distribution covers 68 percent, 95 percent, and 99.7 percent of data. Z is equal to 3^*C_p.

$$Z_u = \frac{USL - \overline{X}}{\sigma} \qquad Z_l = \frac{\overline{X} - LSL}{\sigma}$$

CONDUCT A PROCESS CAPABILITY STUDY: SHORT-TERM (ST) VS. LONG-TERM (LT) PERFORMANCE

A subset of data typically has a smaller variation than the whole dataset. The data are correspondingly referred to as ST vs. LT, which are relative terms. A day is short term compared to a week, but it is long term when compared to an hour. As such, statistical software typically treats a dataset as long-term data.

- ST notations: C_p, C_{pk} (capability entitlement), and PPM_{within} (parts per million)

- LT notations: P_p, P_{pk} (performance under induced noise, same calculation to C_p and C_{pk} except using LT instead of ST standard deviation), and $PPM_{overall}$.

CONDUCT A PROCESS CAPABILITY STUDY: CAPABILITY VS. PERFORMANCE

Process capability is the "entitlement" of the process when it is free from the LT effects of variation.

- C_p = on-target performance, not affected by LT variations
- C_{pk} = off-target performance, not affected by LT variations

Process performance represented by P_p and P_{pk} is the process's "real" performance over time when LT variations are induced. When no data are available, the typical default LT shift (or variation) is 1.5 sigma.

Key Takeaways

- Capability analysis is performed to evaluate process performance against specifications to determine if a process is capable of meeting customer requirements.
- Key measures of process capability include C_p and C_{pk} for ST, and P_p and P_{pk} for LT.
- Higher capability indices indicate better process performance and better ability to meet customer needs. $C_{pk} > 1.3$ is usually expected.
- Lower capability identifies opportunities for process improvements.

Chapter 13
Hypothesis Testing and ANOVA

While SPC, MSA, and capability analysis are the three signature LSS analyses, general statistical analyses are also powerful data analytical tools to support LSS projects. *Hypothesis testing* is an essential tool to assess the statistical significance of observed changes or differences and test assumptions about processes. *Analysis of variance* (ANOVA) is a fundamental statistical method and the basis (core analysis) of many statistical analyses.

Hypothesis testing is a statistical method for testing assumptions about a population parameter based on sample data. It involves comparing a null hypothesis (H0) against an alternative hypothesis (H1) to determine if there is enough evidence to reject the null hypothesis. A few definitions are listed below.

- **Null hypothesis (H0):** A statement that there is no effect or difference, often representing the status quo.

- **Alternative hypothesis (H1):** A statement that there is an effect or difference.

- **Significance level (α):** The probability threshold for rejecting the null hypothesis, commonly set at 0.05.

- **P-value:** The probability *(p)* of observing the data or something more extreme if the null hypothesis is true. We will reject the null hypothesis if $p < \alpha$.

- **Test statistic:** A value (for example, *t* or *F*) calculated from the sample data, used to determine whether to reject the null hypothesis.

TYPES OF HYPOTHESIS TESTS

See Figure 13.1 for a visual of these four tests:

1. **T-test:** Compares the means of two groups or the mean of a group against a known value.

2. **F-test:** Compares the variances of two samples or the ratio of variances between multiple samples.

3. **Chi-square test:** Tests the relationships between categorical variables.

4. **Regression analysis:** Tests the relationships between continuous variables.

		X Input	
		Discrete	Continuous
Y Output	Discrete	Chi-square	F-test
	Continuous	T-tests	Regression

Figure 13.1 Hypothesis tests.

ANALYSIS OF VARIANCE (ANOVA)

ANOVA is used to compare the means of two or more groups to determine if at least one group's mean is significantly different from the others. It helps identify whether the observed variations between groups are due to real differences or random variation.

- One-way ANOVA compares the means of two or more independent groups based on one factor.

- Two-way ANOVA involves two factors and can assess both the main effects and the interaction effect between the factors.

Components of ANOVA

- *Total variation* is shown as $SS_{total} = SS_{factor} + SS_{error}$.

- *Between-group variation* measures differences between group means. SS_{factor} for all, or MS_{factor} per degree-of-freedom (DF).

- *Within-group variation* measures variation within each group. SS_{error} for all, or MS_{error} per DF.

- *F-statistic* is the ratio of between-group variability to within-group variability = MS_{factor}/MS_{error}. A large F-statistic suggests that the differences between group means are significant.

ANOVA Example:

Source	DF	Seq SS	Adj SS	Adj MS	F	P
Main effects	3	73.111	73.1111	24.3704	8339.44	0.008
Two-way interactions	3	70.333	70.3329	23.4443	8022.55	0.008
Residual error	1	0.003	0.0029	0.0029		
Total	7	143.447				

In this example, both the main effect and interactions are statistically significant, since $p < .05$.

Applications of ANOVA

- Quality control: ANOVA can identify which machines, shifts, or processes contribute most to variability in product quality.

- Process improvement: You can prioritize areas for improvement by identifying significant factors influencing process performance.

- Experimental design: ANOVA is essential for understanding the effects of different experimental conditions.

Key Takeaways

- Hypothesis testing and ANOVA are critical tools in LSS for making data-driven decisions and validating improvements.

- These methods help determine the statistical significance of differences observed in data, guiding actionable insights and interventions.

Chapter 14
Correlation and Regression

Correlation and regression analysis are fundamental statistical techniques to explore and quantify the strength and direction of relationships between variables, identify key predictors, and develop predictive models. They can be seen as an extension of ANOVA, going from discrete input variables to continuous input variables.

CORRELATION ANALYSIS

Correlation measures the strength and direction of the linear relationship between two continuous variables using a value between –1 and 1 called the *correlation coefficient (r)*, as illustrated in the scatter plots in Figure 14.1.

    A. Negative correlation ($r < 0$): As one variable increases, the other decreases.

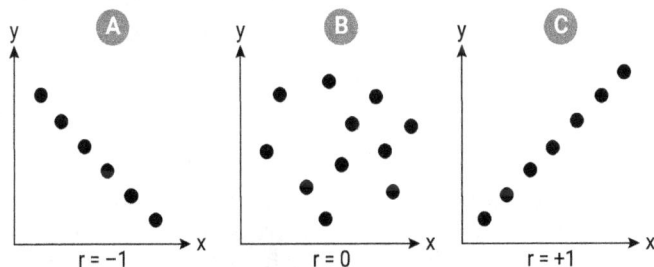

Figure 14.1 Correlation coefficient *(r)*.

B. No correlation ($r \approx 0$): No linear relationship between the variables.

C. Positive correlation ($r > 0$): As one variable increases, the other also increases.

REGRESSION ANALYSIS

Regression analysis estimates the relationship between a dependent variable (outcome) and one or more independent variables (predictors). Types of regression include:

- Simple linear regression, which examines the relationship between one dependent variable and one independent variable.

- Multiple regression, which analyzes the relationship between one dependent variable and two or more independent variables.

The regression equation is the formula representing the relationship, typically in the form $Y = a + bX$, for simple linear regression, where:

- **Y** = Estimated dependent variable (outcome)

- **a** = Constant—Y-intercept (value of Y when X = 0)

- **b** = Estimated slope (change in Y for a one-unit change in X)

- **X** = Independent variable (predictor)

COMPONENTS AND METRICS

- *Coefficient of determination (R^2)* indicates the proportion of variance in the dependent variable explained by the independent variable(s). $R^2 = r^2$, ranging from 0 to 100 percent. Greater than 90 percent indicates a good/strong prediction capability, and greater than 60 percent indicates a usable model.

- *P-value* determines the statistical significance of the regression coefficients. A *p*-value less than the significance level (α, typically .05) suggests a significant relationship.

- *Residuals,* differences (errors) between observed and predicted values of the dependent variable, are used to assess the fit of the regression model.

Regression Example:

The regression equation is   Strength = 2628 – 37.2 Age

Predictor	Coef	StDev	T	P
Constant	2627.82	44.18	59.47	0.000
Age	–37.154	2.889	–12.86	0.000

S = 96.11   R–Sq = 90.2%   R–Sq(adj) = 89.6%

Analysis of Variance

Source	DF	SS	MS	F	P
Regression	1	1527483	1527483	165.38	0.000
Residual Error	18	166255	9236		
Total	19	1693738			

In this example, both constant and age are statistically significant since both $p < .05$; the model is strong since R–Sq > 90 percent.

Practical Applications

- Quality control: Analyze the relationship between process variables and product quality to identify areas for improvement.

- Process improvement: Identify key variables influencing process performance and optimize them to improve outcomes.

- Predictive analytics: Develop models to predict future outcomes based on historical data, aiding in decision-making and planning.

Key Takeaways

- Correlation and regression analysis are powerful tools for exploring relationships between variables and making predictions.

- Correlation measures the strength and direction of a linear relationship, while regression quantifies the relationship and allows for predictions.

- Key metrics in regression analysis include the correlation coefficient (r), coefficient of determination (R^2), and p-values for significance testing.

Chapter 15
Basic Design of Experiments (DOE)

Design of experiments (DOE) is powerful in identifying and verifying improvements, by quantifying the relationship between key process input variables (KPIV) and output variables (KPOV). Primarily used in the Improve phase, DOE allows us to systematically and efficiently identify the most significant factors affecting our processes and optimize them to meet customer requirements. We will wrap up the data analytical coverage with DOE before kicking off the Improve phase.

VARIOUS TYPES OF EXPERIMENTS

- One-factor-at-a-time (OFAT) testing is an improvement to T&E, intuitive yet still inefficient, and is commonly used by engineering. A serious drawback is it cannot estimate interactions.

- Designed experimentation (DOE) experiments with multiple changes (factors) at a time through a strategically designed structure. It is used:

  - To test multiple changes at the same time but study their impacts separately.

  - To gain efficiency; data are shared and reused by all factors via different combinations, each having multiple data points via different combinations of data.

  - To gain effectiveness—for example, with interaction effects.

  - To provide data well-suited for statistical analysis.

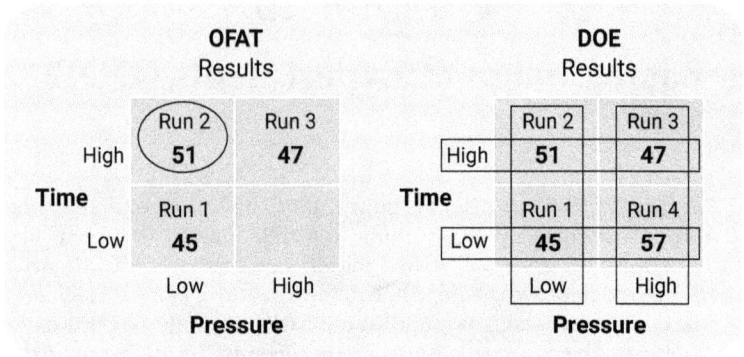

Figure 15.1 OFAT vs. DOE illustration.

In Figure 15.1, OFAT has a single data point for each change, with no visibility on effect when changing time and pressure together. DOE has two data points for each change and visibility on effect when changing time and pressure together, effectively doubling the ROI.

POPULAR TYPES OF DOE

- Full factorials test all combinations of factors, including interactions; they provide complete information about the factors, but are expensive in terms of the number of runs required.

- Fractional factorials test a fraction of the possible combinations of the factors; they may deliver only the required portion of information, which is more efficient than a full factorial.

- Response surface designs provide a detailed "topographic map" of the response against factors; these are used for fine-tuning factor levels for optimum response.

- Mixture experiments are used in situations where the factors (ingredients) must sum up to 100 percent, such as chemical-related applications.

THE CONCEPT OF SEQUENTIAL DOE

Even though DOE is more efficient than OFAT, it's still not efficient enough. Instead of doing a single, catch-all DOE, a more common practice is to work sequential DOE, starting broad and shallow, and then narrowing the field while going deeper in the following stages, as illustrated in Figure 15.2. As an example, you may complete:

- A screening DOE that narrows the field of variables under assessment

- A full-factorial DOE that studies the response of every combination of narrowed factors and factor levels with an attempt to zone in on a sweet spot

- A response surface DOE that optimizes the response in the narrowed sweet spot

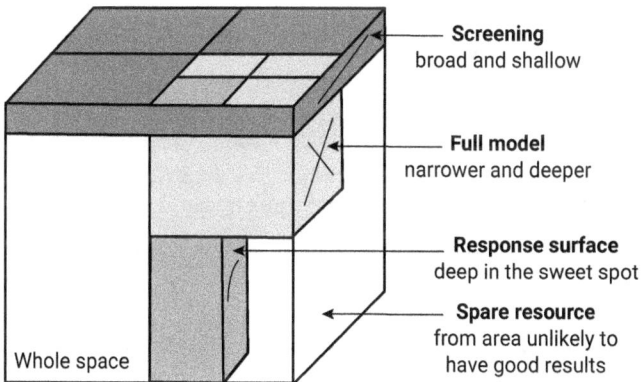

Figure 15.2 Sequential DOE concept.

Terms and Definitions

- **Response:** A measurable output of the system, typically a KPOV. It can be quantitative (for example, temperature and RPM) or qualitative (for example, good, better, and best). This chapter will only cover quantitative responses.

- **Inputs:** Variables that affect the responses, such as voltage, velocity, position, raw materials, etc.

- **Control factors:** Inputs that affect the response and are actively controlled and changed in DOE, such as resistor values, diameter, material property, etc.

- **Noise factors:** Inputs that affect the response but are not actively controlled and changed in DOE, such as ambient temperature, humidity, altitude, load, etc.

A FEW NOTES

- A full factorial is known as an i^k design – (#levels)$^{\text{#factors}}$, where k is the number of equal-level factors and i is the number of levels of the factors.

- If the factors have a different number of levels, it is called a *mixed-level design*. For example, 2 x 4 x 3 refers to three factors, each having two, four, and three levels.

- A treatment combination (tc) is a unique combination of factor levels. For a full factorial, #tc = (#levels)$^{\text{#factors}}$. For example, $2^3 = 8$.

- A run is any combination of factor levels. #runs = (#tc) (#replicates). For example, $2^3(1) = 8$.

- For a full set of "effects" (or "term") of 2^3 design, there are:
  - 3x main effects: A, B, and C
  - 3x two-way interactions: AB, AC, and BC
  - 1x three-way interaction: ABC

- Each factor accounts for (#levels–1) degree of freedom (DF), for example, for 2^3 design:
  - 3x main effects: $1 + 1 + 1 = 3$
  - 3x two-way interactions: $1 \times 1 + 1 \times 1 + 1 \times 1 = 3$
  - 1x three-way interaction: $1 \times 1 \times 1 = 1$
  - Total DF $= 7$

A FULL FACTORIAL EXAMPLE: 2^3 DOE

The 2^3 design is also referred to as L8, which is a popular, versatile, and easy-to-do structure. 2^k DOE can be built up by repeating the low-high setting of each factor, with each new factor added by repeating the previous contrast structure at the low and high setting of the new factor, such as the 2^3 shown in Table 15.1.

Table 15.1 L8 – 2^3 DOE example.

Run	A	B	C	A*B	A*C	B*C	A*B*C	Y
1	−1	−1	−1	1	1	1	−1	44.12
2	1	−1	−1	−1	−1	1	1	47.99
3	−1	1	−1	−1	1	−1	1	53.85
4	1	1	−1	1	−1	−1	−1	45.93
5	−1	−1	1	1	−1	−1	1	47.99
6	1	−1	1	−1	1	−1	−1	52.01
7	−1	1	1	−1	−1	1	−1	58.09
8	1	1	1	1	1	1	1	50.17
+	49.02	52.00	52.06	47.05	50.04	50.08	50.00	Avg = 50.02
−	51.00	48.02	47.96	52.97	50.00	49.94	50.04	
Effect	-1.98	3.98	4.10	-5.92	0.04	0.14	-0.04	
½ Effect	-0.99	1.99	2.05	-2.96	0.02	0.07	-0.02	

The interaction terms are found by multiplying the associated factors. Note for 2^3:

1. If omitting (or ignoring) factor C, this 2^3 becomes a 2 x 2 design with two replicates.

2. If replacing the three-way interaction by another factor D, this three-factor full factorial design becomes a half-fractional design for four factors.

ANALYSIS OF MEAN (ANOM)

The effect of a factor (or term) is the difference in the response between the two levels of the term. A half-effect is the sensitivity of the response to a unit change of a term. We can build a regression-like transfer function using the half-effects and the grand average. $Y = 50.02 - 0.99A + 1.99B + 2.05C - 2.96AB + 0.02AC + 0.07BC - 0.02ABC$.

This is the formula to estimate the result (response) based on the input variables (factors). However, we do not know which, if any, of these effects are statistically significant, which is what DOE analysis will do (via software, which is not covered in this book).

Key Takeaways

- DOE can systematically and efficiently identify and optimize the most significant factors (KPIV) to achieve the desired and optimized outcomes to meet customer requirements.

- Key components of DOE include factors, levels, responses, and experimental design.

- Common types of DOE include fractional factorial, full factorial, and response surface designs.

- Sequential DOE uses the aforementioned processes in a progressional way to achieve higher efficiency.

Improve Phase

One of the focuses of the LSS project is to identify a handful of vital few KPIVs that affect KPOVs the most during Measure-Analysis iterations. With that finding, the project can advance to the Improve phase.

Typical activities in the Improve phase include ideation and validation, which involve piloting the ideas and data analytics. These activities are embedded in the individual tools covered for this phase.

Traditionally, people tend to introduce tools in the phase in which they are primarily used, which means DOE is typically introduced in the Improve phase. As explained earlier, we cover all statistical analyses in the Analysis phase, although they may be used in other phases.

On the other hand, while lean activities can be used in other phases, they add the most value in the Improve phase; thus, we will dedicate the Improve phase to the seven most popular lean tools and applications, starting with kaizen.

Chapter 16
Kaizen Methodology

The Japanese methodology for "improvement" or "change for the better" refers to the philosophy and practices of continuous improvement (CI). It breaks apart the current state, quickly analyzes it, and puts it back to make it better. There are five kaizen principles:

1. Know your customer

2. Let it flow

3. Go to gemba (or the real place)

4. Empower people

5. Be transparent

TYPES OF KAIZEN

- Spot or point kaizen: Maximum of one week of preparation with one to two days of implementation, a small team of three to five people, focusing on a single workplace with little money spent.

- Kaizen event: Maximum of four to six weeks of preparation with three to a maximum of five consecutive days of implementation, five to seven people participating in the event week with Yellow Belt (non-statistical) tools applied, to reduce waste and improve area KPIs with little or no money spent.

- Kaizen project: Requires a longer Measure phase to gather appropriate data for analysis, with a goal of significantly improving VSM KPIs by applying Green/Black Belt tools. Execution takes several weeks, not requiring full-time participation of the team. May involve significant investments to realize savings.

KAIZEN PREPARATION AND EXECUTION

To charter the event and team, review the implementation plan that was generated from diagnostic mapping/VSM and finalize the target area and tools to be deployed; collect the needed data; identify a team to include the needed functions and expertise; and work with the scheduling group to provide a production window to conduct the event.

Here is a typical daily cadence:

- The sponsor hosts a kick-off to state the goals and leadership commitment.

- A GO-team meeting starts each day, to quickly review the work done on the previous day.

- A progress review meeting ends each day, to brief the sponsor.

- A small celebration of the achievements is organized at the end.

Detailed Schedule Example

Day 1: Kick-off. The sponsor gives a brief overview of the goal.

Day 1: Team training. The facilitator (or event leader) provides training on pertained tools.

Day 1: Current state.

- Collect data. Certain historical data should be collected and analyzed prior to the event. Additional data may be identified and collected during the event.

- Visualize the current condition. Identify inefficiencies and waste. Document opportunities.

Day 2: Future state.

- Brainstorm ideas. "Bleed the team dry."

- Classify ideas into four categories: do now, do within two weeks, do longer than two weeks, and don't do.

- Identify and visualize ideal future-state conditions. Develop desired targets.

Day 2: Planning. Develop a detailed improvement plan.

- Keep the team focused on the goal (reduce inventory or setup time, develop a cell, etc.).
- Assign tasks to small teams. Prepare for implementation.

Days 3 and 4: Implementation.

- Complete implementation. Evaluate changes made.
- Develop control plans and standard work to sustain.
- Capture all the actions that were not 100 percent completed.
- Prepare training records for HR and certificates for the team.

Day 5: Report out.

- Prepare report out.
- Report out to the site leadership team (15 minutes), followed by a final walkthrough of the area.

Post-event follow-up: Conduct a weekly review meeting for the next four weeks or until all action items are closed. Update the implementation plan as needed.

Point Kaizen

Examples of point kaizen include eliminating waste, variation, over-burdened points, and/or unreasonable conditions in the process. Point kaizen principles include:

- All work at belly-button height
- Horizontal work with no blind operations
- Part presentation span at 45° fan and arm reach of 12 inches
- All assembly/part/machines at a common height
- Parts and tools in the order of use
- Easy-to-read/select/grasp parts and tools
- Natural posture of work (resembling walking position)

Key Takeaways

- Kaizen methodology is a key principle and technique of lean focusing on continuous improvement.

- It can be done alone or embedded in LSS projects, particularly in the Improve phase.

- It can be in a formal week-long event or in a less formal ad hoc format.

Chapter 17
5S Visual Workplace

$5S$ is a foundational tool in lean focusing on organizing the workplace for efficiency and effectiveness. It's usually one of the very first items in a lean initiative or any improvement effort. It is worth noting that most lean techniques/tools support each other. Although each has a different focus, they are all derived from the same lean principles. For example, 5S and visual workplace help each other where processes are visualized and streamlined, waste is minimized, and productivity is optimized.

5S

5S is a methodology that drives a workplace to be clean, uncluttered, safe, and well organized to help reduce waste and optimize productivity. It's designed to help build a quality work environment, both physically and mentally. Sometimes safety is added to become 6S.

- S-1 **Sort:** Sort needed and unneeded items from the area.

- S-2 **Store:** Set in order. Organize places to store the needed.

- S-3 **Shine:** Keep everything clean.

- S-4 **Standardize:** Define standards to maintain an organized workplace.

- S-5 **Sustain:** Monitor and commit 5S to become a culture of daily discipline.

VISUAL WORKPLACE

A visual workplace is a self-ordering and self-regulating work environment that uses visual signals or displays to convey self-explaining information to all who come within the visible range. It communicates important information, makes waste stand out, and drives actions.

Here are the differences between visual display and visual control:

- A visual display shows history and status, provides information, and maintains status quo.

- Visual control provides needed current information that can be acted upon; it displays abnormalities to alert everyone to provide fail-safe processes.

REASONS FOR A VISUAL WORKPLACE

Humans record information as follows: 83 percent by eyesight, 11 percent via hearing, 3.5 percent via smell, 1.5 percent by touching, and 1 percent by tasting. And humans retain information as follows: 20 percent of what is heard, 30 percent of what is seen, 50 percent of what is heard and seen, 70 percent of the subjects that are discussed, and 90 percent of what is personally experienced.

Key Takeaways

- 5S visual workplace is a foundational tool in lean, focusing on organizing the workplace for efficiency and effectiveness.

- 5S builds on the principles of kaizen by creating a visual workplace that supports and sustains continuous improvement.

- The 5S principles include sort, set in order, shine, standardize, and sustain.

Chapter 18
Mistake Proofing

A nother popular lean technique is *mistake proofing* to prevent errors and defects from occurring, leading to higher-quality products and services.

Poka-yoke (ポカヨケ), the Japanese term for mistake proofing, refers to any mechanism that helps an operator avoid *(yokeru)* mistakes *(poka)*. Its purpose is to eliminate product defects by preventing, correcting, or drawing attention to human errors as they occur. It's better to prevent an error than to correct one. Early detection at the source is less expensive and avoids the bullwhip effect.

Source, self, and successive checks make mistake proofing much more effective than other detection methods. Build 100 percent inspection into the process with feedback and the corrective action loop as short as possible. Effectiveness varies for various poka-yoke practices:

- End-of-line inspection = poor method
- In-process inspection = good method
- Self-check certified workers = better method
- Certified processes = best method

FOUR METHODS OF POKA-YOKE

Following are four methods (listed from strong to weak):

- **Control:** Eliminate the opportunity for errors to occur. For example:
  - Regulatory functions (warning and control method)
  - Two-sided car keys with no wrong insertion
  - Tank connections for grills

- **Shutdown:** Stop the process when the error occurs. This is one of the four basic elements of Jidoka (automation with a human touch): detection, stoppage, response, and prevention.

- **Warning:** Use visual or audio signals to notify operators or users when an error occurs. It's better to install mistake-proofing devices inside the process to provide feedback on errors immediately. For example, you may use error messages to enable operators to take quick corrective actions.

- **Sensory:** Make the error obvious, but leave it to users to perform the check. Smart Gate with infrared sensing devices is a low-cost sensor system example that is primarily used for large parts (castings, cartons, boxes, or literature) or single bin applications.

Common poka-yoke devices and techniques include physical devices, and procedural and behavioral controls, such as sensors, templates, interference pins, limit switches, proximity sensors, vision systems, odd part kitting, counters, color coding, sequence restriction, check fixtures, symmetry, and asymmetry. (See Figure 18.1.)

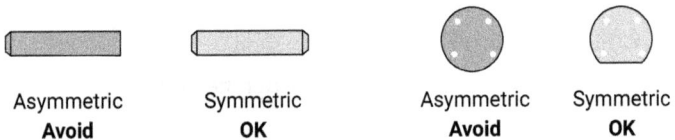

| Asymmetric | Symmetric | Asymmetric | Symmetric |
| **Avoid** | **OK** | **Avoid** | **OK** |

Figure 18.1 Symmetry or exaggerated asymmetry to eliminate orientation problems.

Key Takeaways

- Mistake proofing is a technique used to prevent errors and defects from occurring.

- Early in-process detection is most effective in preventing the bullwhip effect.

- Types of mistake-proofing devices include physical devices and procedural and behavioral controls.

- Symmetry vs. asymmetry is a frequently used technique.

Chapter 19
Standard Work

S*tandard work* (SW) is a foundational concept in lean involving documenting and following the best practices for performing a task to ensure process consistency, efficiency, and quality.

Standard work, or standardized work, is an agreed-upon set of work procedures that establishes the best method and sequence known today for each manufacturing or transactional process. For example, a call center may follow a standard script to interact with customers. Standard work is intended to maximize human and equipment efficiency while simultaneously ensuring safe, ergonomic conditions.

WHY USE STANDARD WORK?

It's the basis for process stability and a baseline for continuous improvement. Taiichi Ohno, the father of the TPS, said, "Without standards, there can be no improvement." If a process is not standardized, what are you improving? Randomness?

Standardization is a very effective means for creating the most consistent performance possible and reducing variation (a major connection between lean and Six Sigma). It is part of the continuous effort to establish accountability and identify problems, driven by people, not done to people. Everyone is responsible for maintaining SW and making it work properly.

SW describes the safest, most ergonomic, and efficient way known today; enables consistent and repeatable work by eliminating waste, limiting WIP, and minimizing mistakes; defines steps that can easily be read and understood by everyone to establish accountability; measures our capability to meet customer demand; and establishes a baseline to identify continuous improvement opportunities.

Notice the different uses of SW and work standards (or work instruction/WI, a set of guidelines that employees follow to complete work tasks and achieve desired results). Leaders use SW, not WI, to evaluate if operators are working according to the known standard; if not, ask why and implement improvements. In short,

- SW is an uncontrolled document for leaders.
- WI is a controlled document for operators.

STANDARD WORK DOCUMENT TYPES

There are various types of standard work documents for different needs. Five options are listed here:

- Standard work production sheets visually identify the sequence of tasks and display three key elements:
  - Takt time, the production rate needed to meet demand
  - The precise work sequence to perform tasks
  - The standard inventory (WIP), including units in machines required to keep the process operating smoothly
- Standard work combination sheets define the time for work-machine interactions.
- Standard work process capacity sheets calculate true capacity and bottlenecks.
- Standard work instruction sheets provide detailed descriptions of each required process step for both operational and business processes.
- Standard work cell sheets visually identify the layout of the cell with the quantity and position of the operators, including material flow and layout of raw materials and finished goods.

TIME OBSERVATION

A key activity in standard work is time observation. Time observation isn't a time study for setting rates; it is a tool to map and document all elements of the process so waste can be identified and reduced or eliminated. Two-person teams are recommended for time observations. One serves as the timer and the other as the recorder. Involve the operators in the time observation, but do not interfere with the operation. Assure the workers that you are NOT measuring them personally but rather the process.

Team up with a couple of people. Confirm the task sequence with the worker. Break down tasks into the smallest measurable elements; combine the very small ones. Write down the steps on the time observation form. Have a stopwatch and pen ready. (Use your cellphone stopwatch app.) Practice three cycles as a rehearsal before beginning timing. Nowadays, time observations are performed via video recording rather than manually recording live sessions in-line. See Chapter 21 for an example of changeover time observation.

Recording Techniques

Record accumulative times for several cycles of operation. The stopwatch does not stop over multiple cycles. This feature is supported by most smartphone stopwatch apps. Calculate the individual task times after the observation is completed. Take multiple time measurements, depending on variability and length of cycle. Complete at least five cycles, but ten is better. *Do not* take the average time of the cycles since this includes waste in process. The lowest repeated time is the correct time to use since it obviously had the least waste during the time observation.

Key Takeaways

- Standard work is the documented best practice at the time for performing a task to ensure consistency, efficiency, and quality in processes.

- Key elements of standard work include the work sequence, standard work cycle time, takt time, and standard WIP.

- Time observation is used to decide best practice.

Chapter 20

Continuous Flow

Continuous flow (CF) is a core lean principle focusing on optimizing the flow of work to minimize waste and improve efficiency. Also known as one-piece flow, it is a production method where products move continuously through the manufacturing process with minimal interruption or waiting times. This maximizes efficiency, reduces waste, and maintains a consistent flow of materials and work while minimizing inventory and working capital.

As shown in Figure 20.1, CF eliminates waiting and inventory between process steps (WIP), allowing the product to flow through the value stream faster and opening up capacity. As a contrast, batch manufacturing takes longer and reduces capacity.

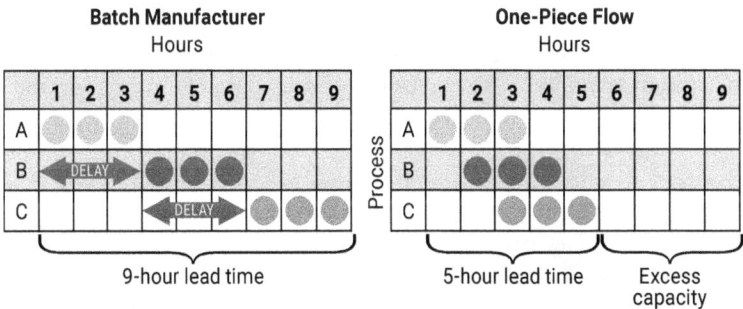

Figure 20.1 Batch flow vs. one-piece flow.

In traditional batch production, customer orders are produced in batches; processes are arranged in functional silos; and supplier processes "push" products to the customer whether needed or not.

Batch production focuses on individual process efficiency (quantity and speed). Work between functional departments is often unbalanced, with higher WIP and stock (including safety stock to accommodate variations in supplier and upstream processes). The production pattern is not synchronized with the pattern of customer demand. The system is inflexible and requires extraordinary effort to schedule and respond to even small changes in demand.

LEAN MANUFACTURING STRATEGIES

There are several strategies that produce a lean operation. Calculate the takt time for the products or services, and ensure all cycle times are either less than or equal to the takt time; this may require many improvement activities (kaizen, Six Sigma). Base the process capacity or equipment capacity on takt time. Make manufacturing layout support one-piece flow as much as possible, or introduce a pull system (including kanban, Heijunka/production leveling, etc.). If you cannot flow, you pull.

Takt Time

From the German word "takt zeit" for "measure time," takt time is the maximum amount of time to produce a product unit to satisfy demand.

$$\text{Takt time} \quad \frac{\text{Available time}}{\text{Customer demand}}$$

Production leveling comes into play when customer demand is too unpredictable. As a result, operations may plan for a fixed quantity of finished goods per cell per day instead of constantly adjusting for takt time fluctuation.

Push vs. Pull

Notice the difference between the two strategies below:

- Push: Make all we can just in case. Typically based on anticipated usages and production approximation; associated with large lots and high inventories.

- Pull: Make what's needed when needed. Typically based on actual consumption and precise production; associated with small lots and low inventories.

Continuous Flow (CF)

Produce and move one item at a time (or a small and consistent batch of items) through a series of processing steps as continuously as possible. Each step makes just what is requested by the next step. CF can be in many forms, from moving assembly lines to manual cells. One-piece flow is the most ideal state to attain.

First-In, First-Out (FIFO) and Supermarket

Flow where you can; pull where you can't. Pull replenishment systems typically involve FIFO storage or supermarkets between process steps to control inventory, which requires kanban to trigger production. A FIFO is for a single part. A supermarket is for multiple parts, each having its FIFO lanes. Upstream production stops when FIFO or the supermarket is full. Typically, a process step near the customer side will serve as a pacemaker for the process to respond to customer demand and control the production pace via kanban.

Key Takeaways

- Continuous flow is a core lean principle focused on optimizing the flow of work to minimize waste and improve efficiency.
- Key components of CF include one-piece flow (the most ideal state), pull systems, and takt time.

Chapter 21
Setup Reduction

Although all lean techniques covered so far support continuous flow, two more tools particularly do: setup reduction and cell design, which we'll discuss next.

Setup reduction, also known as quick changeover or single-minute exchange die (SMED), is the process of reducing the time needed to change over a process from the last part for the previous product to the first good part for the next product.

WHY SETUP REDUCTION?

A lengthy changeover time costs money because people and equipment are not producing. Reducing changeover time increases production's flexibility to support multiple customers' demands for products produced on the same equipment.

Here are some typical changeover approaches:

- Limit the number of changeovers by running large lot sizes. This is usually associated with high carrying costs of excess inventory.

- "Just do it" and perform as many changeovers as needed to meet customer demand. This is usually associated with long changeover downtimes and poor productivity.

- Deploy quick changeover to reduce changeover time. This is the best option with low inventory and improved productivity.

An internal setup must be done when the machine stops, like removing dies and tooling. An external setup can be done when the machine is running, like preparing tooling. The key to minimizing changeover time is to turn an internal setup into an external setup.

FIVE-STEP APPROACH

Follow these steps to reduce setup time.

1. **Measure**

   - Videotape the changeover process, recording at least 7–10 cycles of the changeover process.

   - Observe the process (live or video), documenting each step on the time observation form. Typical steps include:

     - Searching for tools, jig, fixtures, etc.

     - Changing fixtures or tools

     - Adjusting a program or fixture

     - Walking or moving around to retrieve tools, jigs, and fixtures

   - Record task times on the time observation form. This typically requires two people: one to monitor a stopwatch and to call out the times and the other to log results. The recorded times should be accumulative; that is, let the stopwatch run continuously rather than stop and start for each step. After completing all observations, calculate the incremental task times.

2. **Separate**
   Classify each changeover step as internal or external.

3. **Convert**
   Convert as many internal setup steps to external setup steps as possible (called *externalizing*):

   - What has to be done while the equipment is shut down?

   - What can be done while the equipment is running?

Techniques:

- Use a fixture or pallet changer: Install the next part on an external jig or fixture while the machine is running. The change will be only a quick swap of loaded fixtures.

- Prepare: Gather tools and equipment (setup kitting) while the machine is running. Preheat fixtures if applicable and possible (for example, molding reheat fixtures).

- Use pre-checks: Ensure function for all gages, tools, etc. before the changeover begins.

- Consolidate mini-setups: Group separate manual internal elements into one larger manual element to get rid of "prison bars" type of frequent interruptions.

4. **Streamline**
   Streamline both the internal and external steps of the changeover process. Look for the eight wastes.

5. **Standardize**
   Implement standard work to document the updated and improved changeover process.

Key Takeaways

- Setup reduction is a fundamental lean technique focused on reducing the time to change over a process from producing one product to another.

- Setup reduction builds on the principles of continuous flow and standard work by further reducing waste and improving efficiency through minimized changeover time.

- The key benefits of setup reduction include increased flexibility, reduced waste, and improved efficiency.

Chapter 22
Cell Design

Cell design (CD) is the last lean concept covered in this book. Cell design refers to the location and layout of processing steps immediately adjacent to each other so parts, documents, etc. are processed in nearly continuous flow, either one at a time or in small batch sizes throughout the complete sequence of processing steps. The idea is to organize workstations and processes into cells to improve efficiency, quality, and flexibility.

USING CELL DESIGN

- Create product families using a process matrix. Map product family value streams.

- Understand the demand and calculate takt time.

- Capture manual and auto time via time observations.

- Balance the line via the operator cycle-time balance chart.

- Capture cell design and work content with a standard work cell sheet.

- Design the layout of workstations to minimize movement and transportation of materials and products.

- Cross-train employees to perform multiple tasks within the cell to improve flexibility and responsiveness.

Two-Step Approach

Follow these steps to successfully navigate cell design.

1. Identify and eliminate waste

- Capture manual operation time for each operator.

  - Manual internal time: The operator completes while the machine is down to eliminate or reduce via externalization.

  - Manual external time: The operator completes while the machine is running.

- Capture auto time—the cycle time for each machine—to use for line balancing. Use available cycle-time data to calculate a weighted average cycle time for multiple part numbers.

2. Design cell layout

A U-shaped cell has significant advantages over other layouts: clear production status (short/full), reduced floor space, and increased worker flexibility across operations. Thus, it's the most favored layout in lean (see Figure 22.1).

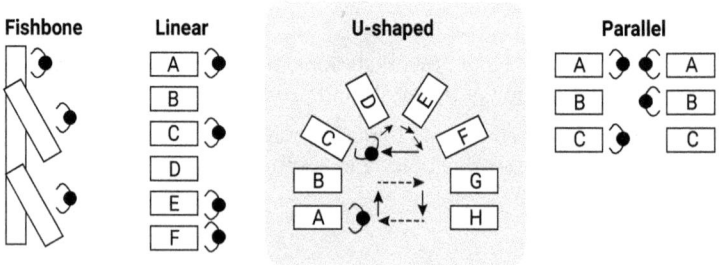

Figure 22.1 Consider cell layout options.

Key Takeaways

- Cell design is a lean concept focused on organizing workstations and processes into cells to improve efficiency, quality, and flexibility.
- Key elements include product families, takt time, line balancing, and cross-training.

Control Phase

After achieving the desired improvements, the LSS project moves into the Control phase. The control plan is a core activity of the Control phase, providing a structured approach to sustain improvements over time.

Other Control phase activities include SPC and updating or finalizing various documents (for example, FMEA and standard work).

It's also the time to develop the project report as part of project wrapups. In practice, the project report is more likely developed throughout the project progression and finalized in this stage; thus, it is formally addressed here as the last topic.

Chapter 23
Control Plans

Consider your strategy or how the control plan is integrated with other activities. The control plan intends to operate consistently on target with minimum variation; minimize process tampering (over-adjustment); ensure implemented improvements are institutionalized; ensure compliance with industry standards (ISO 9001, etc.); and provide adequate training to all procedures including required maintenance schedules.

Components of the control plan include process map steps, short- and long-term capability, key process input variables (KPIVs) with appropriate tolerances and control limits, key process output variables (KPOVs), targets and specs, important noise variables (uncontrollable inputs), and designated control methods including tools and systems. For example, this could include:

- Checklists or mistake-proofing systems
- Statistical process control (SPC)
- Automated process control or workmanship standards
- Standard operating procedures (SOPs)

ROADMAP

Follow these steps to create a control plan.

1. Collect existing documentation for the process.
2. Determine the scope of the process for the current control plan.
3. Form teams to update existing documentation.
4. Replace short-term with long-term capability.

5. Fill in the control plan form.

6. Identify missing or inadequate information.

7. Pay particular attention to adequacy of training, maintenance, and operating and reaction plans.

8. Assign tasks to complete missing elements of the plan.

9. Verify compliance with plant documentation.

10. (Re)train operations personnel as required.

11. Collect required sign-off signatures.

12. Verify control plan effectiveness in six months.

FMEA VS. CONTROL PLAN (STEP 5)

FMEA should be a primary source for identifying key variables to control and an initial review of the current control plan. A control plan itemizes more details of control methods. For example, you may use a control plan for:

- Continuous/discrete variables
  - Inspection and test methods, SOPs, checklists
  - Mistake-proofing systems and devices: color tags, single-use devices
  - SPC: choice of control chart and sampling plan, including procedures for identifying and responding to out-of-control points and undesirable trends, and preventing the recurrence of special causes
  - Automated process control
- Noise (uncontrollable input) variables
  - Compensation for changes in KPIVs (based on data or known relationships, not opinion)
  - Shutdown conditions
  - Alternative quality control procedures (inspection, sorting, rework, etc.)

Key Takeaways

- A control plan is a core component in the Control phase of LSS methodology, providing a structured approach to maintain process performance and ensure improvements are sustained over time.

- Key components of control plans include process steps, key process parameters, measurement methods, and response plans for out-of-control conditions.

Chapter 24

Project Report and Presentation Skills

Presentation skills for effective communication are essential in LSS projects to ensure project teams can convey their findings, recommendations, and progress to stakeholders. You'll need these skills throughout every step of the project, starting at the very beginning—but we'll formally discuss presentation and the project report here.

PRESENTATION SUITED FOR BOTH EXECUTIVES AND DOERS

Engineering presentations have some unique needs. Note that different types of presentations have different setups. For the general public, I suggest listing only key points on the slides, in a 5 x 5 format (five bullet points per slide and five words per bullet). Details are considered distractions and can be provided verbally. Yet engineering presentations do need details and depth to allow listeners to dig in offline. Finding a good balance between bones (for executives) and meat (for doers, like engineers) will add great value.

Here are some tips for presentations suited for both executives and "doers" (engineers):

- Make sure the storyline flows and is easy to follow. Make the connections between slides as straightforward and obvious as possible.

- Each slide considers full bone-meat-skin structure:

  - Bone: Include a summary line in each slide to highlight the key message (theme, executive highlights) per slide, in a consistent, eye-catching format (bold, box with colored background, etc.).

- Meat: Provide supporting evidence (data) for people (like engineers) who want to know details for research off-line. Highlight the key points using a consistent method (circle, color, comments, etc.)

- Skin (format): Use an appealing format/interface. A picture is worth a thousand words.

THE STORY ARC

A story arc (also called dramatic arc or narrative arc) is a great structure for LSS presentations to follow. See Figure 24.1 for a common story arc that proves useful in DMAIC presentations.

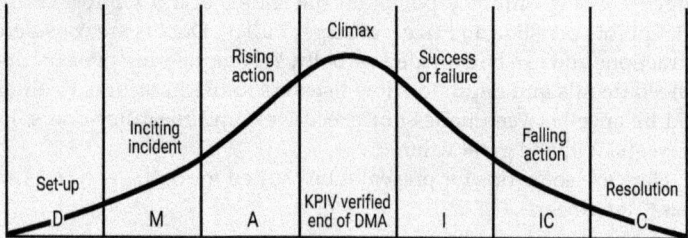

Figure 24.1 Story arc aligned with DMAIC.

Key Takeaways

- Effective communication is essential in LSS projects to ensure project teams can convey their findings, recommendations, and progress to stakeholders.

- Make the connections between slides as straightforward and obvious as possible.

- Consider full bone-meat-skin structure for each slide.

- A narrative arc is a great structure for LSS presentations to follow.

Acronyms and Symbols

3P—Production Preparation Process

5S—Sort, Store, Shine, Standardize, Sustain

8D—The 8-discipline problem-solving approach

A3—The Toyota practice of documenting the problem solving on an A3 (11 x 17-inch) paper

ANOM—Analysis of Mean

ANOVA—Analysis of Variance

BB—Black Belt

CC—Control Chart

CD—Cell Design

CDOV—Concept, Design, Optimize, Verify

C-E Matrix—Cause-and-Effect Matrix

CF—Continuous Flow

C_p—Short-term Process Capability—Centered. Entitlement performance.

CPI—Continuous Process Improvement

C_{pk}—Short-term Process Capability—One-side

CTC—Critical to Customer

CTD—Critical to Delivery

CTQ—Critical to Quality

CTX—Critical to Something

DFSS—Design for Six Sigma

DPM—Diagnostic Process Mapping

DMADV—Define, Measure, Analyze, Design, Verify

DMAIC—Define, Measure, Analyze, Improve, Control

DOE—Design of Experiments

DOWNTIME—Eight wastes

ECO—Easy, Common, Old

FIFO—First In, First Out

FM—Failure Mode

FMEA—Failure Mode and Effects Analysis

GB—Green Belt

GR&R—Gage R&R

H0—Null Hypothesis

H1—Alternative Hypothesis

I-Chart—Individuals Chart

KPI—Key Performance Indicator

KPIV—Key Process Input Variable

KPOV—Key Process Output Variable

LCL—Lower Control Limits

LDFSS—Lean Design for Six Sigma

LSL—Lower Spec Limits

LSS—Lean Six Sigma

MBB—Master Black Belt

MP—Mistake Proofing, poka-yoke

MSA—Measurement System Analysis

NUD—New, Unique, Difficult

NVA—Non-Value Added

OFAT—One-factor-at-a-time

P/T Ratio—Precision to Tolerance Ratio, MSA index

P/TV Ratio—Total variation consumed by standard deviation, GR&R index

PDCA—Plan, Do, Check, Act improvement cycle

P_p—Long-term Process Capability—Centered

PPA—Potential Problem Analysis

P_{pk}—Long-term Process Capability—One-side

r—Correlation Coefficient

R^2—Coefficient of Determination

RACI—Responsible, Accountable, Consulted, Informed

RC—Root Cause

RCA—Root Cause Analysis

RCT—Root Cause Tree

ROI—Return On Investment

RPN—Risk Priority Number

S—Estimated sample standard deviations

SIPOC—Supplier-Input-Process-Output-Customer

SMART—Specific, Measurable, Attainable, Realistic, Time-bound

SMED—Single-Minute Exchange Die

SPC—Statistical Process Control

SR—Setup Reduction

SS—Six Sigma

SW—Standard Work

TIMWOODS—Eight wastes

U—Uncontrolled/uncontrollable Inputs

UCL—Upper Control Limits

USL—Upper Specification Limits

VA—Value Added

VOC—Voice of the Customer

VOS—Voice of the Stakeholder

VSM—Value Stream Map

WIIFM—What's in it for me

X-Y Matrix—Cause-and-Effect Matrix

YB—Yellow Belt

$\bar{x}$—Sample mean

$\bar{s}$—Sample standard deviations

σ—Standard deviations

σ^2_{rpd}—Reproducibility

σ^2_{rpt}—Repeatability

μ—Population mean

α—Significant level

References

Erdil, N.O. and G. Jing. 2022. P21. Lean Six Sigma, *Engineering Management Handbook*, 251-264.

Jing, G. 2008. "Digging for the Root Cause," *Six Sigma Forum Magazine*, May, 19-24.

Jing, G. 2008. "Flip the Switch-Root Cause Analysis Can Shine the Spotlight on the Origin of a Problem," *Quality Progress*, October, 50-55.

Jing, G. 2009. "A Lean Six Sigma Breakthrough-Tiered, Mutually Exclusive Approach," *Quality Progress*, May, 24-31.

Jing, G. 2010. "The Truth About Design for Six Sigma," *Six Sigma Forum Magazine*, May, 26-28.

Jing, G. 2010. "Guest Editorial: Customization is Critical in Deployments," *Six Sigma Forum Magazine*, November, 6-9.

Jing, G. 2011. "Avoid the Pitfalls of VOC," *ISixSigma*, January-February.

Jing, G. 2012. "VOC Should Be VOS," *Six Sigma Forum Magazine*, February, 27-29.

Jing, G. 2014. "FMEA Dilemma," *Six Sigma Forum Magazine*, May.

Jing, G. 2015. Root Cause Analysis Troubleshooting Techniques, *Encyclopedia of Information Systems and Technology*, CRC Press, 1-16

Jing, G. 2015. "Guest Editorial: Is This a Six Sigma Project or Kaizen Event," *Six Sigma Forum Magazine*, February, 5-9.

Jing, G. 2018. "How to Measure Test Repeatability When Stability and Constant Variance Are Not Observed," *International Journal of Metrology and Quality Engineering* 9(10): 1-9.

Jing, G. 2019. "Solve Your FMEA Frustrations," *Lean & Six Sigma Review,* February, 8-13.

Jing, G. 2019. "A Fundamental FMEA Flaw," *Quality Progress,* May, 26-33.

Jing, G. 2022. "Pitch LSS to Management," *Quality Progress,* January.

Jing, G. 2023. "The Missing Ingredient in Building Lean Culture," *Quality Progress,* May.

Jing, G. 2024. "FMEA-lite," *Quality Progress,* May.

Jing, G. 2024. "FMEA Intro," *Quality Progress,* August.

Jing, G. 2024. "A Simplified Approach to FMEA Improves Efficiency," *Lean & Six Sigma Review,* August.

Jing, G. 2025. "Lighten Up—A Tiered Triage Option for FMEA to Deal With Lack of Data in High-mix-low-volume Situation," *Quality Progress,* January.

Liker, J., and D. Meir. 2005. *The Toyota Way Fieldbook,* McGraw Hill.

Mcshane-Vaughn, M. 2022. *The ASQ Certified Six Sigma Black Belt Handbook,* 4th edition, Quality Press.

Munro, R. G. Ramu, and D. Zrymiak. 2022. *The ASQ Certified Six Sigma Green Belt Handbook,* 3rd edition, Quality Press.

NIST/SEMATECH *Engineering Statistics Handbook,* https://www.itl.nist.gov/div898/handbook/

Ramu, G. 2022. *The ASQ Certified Six Sigma Yellow Belt Handbook,* 2nd edition, Quality Press.

Schwartz, B. 2005. *The Paradox of Choice—Why More is Less,* Harper Perennial.

Womack, J., D. Jones, and D. Roos. 2007. *The Machine That Changed the World,* Free Press.